LASERS

Humanity's Magic Light

These and other books are included in the
Encyclopedia of Discovery and Invention
series:

Airplanes: The Lure of Flight
Atoms: Building Blocks of Matter
Computers: Mechanical Minds
Gravity: The Universal Force
Lasers: Humanity's Magic Light
Printing Press: Ideas into Type
Radar: The Silent Detector
Television: Electronic Pictures

LASERS

Humanity's Magic Light

by DON NARDO

The ENCYCLOPEDIA of

D·I·S·C·O·V·E·R·Y
and INVENTION

P.O. Box 289011 SAN DIEGO, CA 92128-9011

Library of Congress Cataloging-in-Publication Data

Nardo, Don, 1947–
    Lasers: humanity's magic light.

    (The Encyclopedia of discovery and invention)
    Includes bibliographical references and index.
    Summary: Discusses the history of lasers and their
uses in such fields as medicine, entertainment, and
the military.
    1. Lasers—Juvenile Literature.   (1. Lasers)
I. Title. II. Series.
TA1682.N37 1990   621.36'6   90-6269
ISBN 1-56006-200-2

Contents

■ ■

Foreword   7

Introduction   10

CHAPTER 1 ■ Pre-Laser Ideas and Inventions   12
The invention of the photophone;
Einstein and stimulated light;
Experimenting with microwaves;
The maser sets the stage for the laser.

CHAPTER 2 ■ The Laser   20
The laser's many inventors;
Gould encounters problems;
The laser is first built;
The birth of the laser era.

CHAPTER 3 ■ The Laser and the Military   27
The attempt to build death rays;
Real battlefield lasers;
The submarine laser;
The future of laser weapons.

CHAPTER 4 ■ The Laser Toolbox   34
Toolbox technology;
Drilling with light;
Burning holes with lasers;
Lasers that weld and cut;
The laser as detective.

CHAPTER 5 ■ The Laser That Can Read and Write   42
Lasers that can scan type;
Supermarket lasers;
Machine voices;
Lasers in the office;
The beam with a memory;
Computer playback;
Computing at light speed.

CHAPTER 6 ■ The Light That Talks   51
  Signals through the air;
  The coming of fiber optics;
  Conversing by light beam;
  The fiber-optic society.

CHAPTER 7 ■ The Laser and Medicine   59
  Bloodless surgery;
  Cleaning arteries with light;
  Restoring the miracle of sight;
  Treating birthmarks;
  The age of painless dentistry.

CHAPTER 8 ■ The Laser and Entertainment   67
  Painting with light;
  Lasers with a beat;
  The laser disc revolution;
  Three-dimensional images;
  Motion picture magic.

CHAPTER 9 ■ The Future of the Laser   76
  An explosion of information;
  Images in thin air;
  A portable appliance store;
  Fixing brains and eyes;
  Providing unlimited energy;
  The power of the atom;
  A world transformed.

Glossary   87
For Further Reading   90
Works Consulted   91
Index   92
About the Author   95
Picture Credits   96

Foreword

The belief in progress has been one of the dominant forces in Western Civilization from the Scientific Revolution of the seventeenth century to the present. Embodied in the idea of progress is the conviction that each generation will be better off than the one that preceded it. Eventually, all peoples will benefit from and share in this better world. R. R. Palmer, in his *History of the Modern World*, calls this belief in progress "a kind of nonreligious faith that the conditions of human life" will continually improve as time goes on.

For over a thousand years prior to the seventeenth century, science had progressed little. Inquiry was largely discouraged, and experimentation, almost nonexistent. As a result, science became regressive and discovery was ignored. Benjamin Farrington, a historian of science, characterized it this way: "Science had failed to become a real force in the life of society. Instead there had arisen a conception of science as a cycle of liberal studies for a privileged minority. Science ceased to be a means of transforming the conditions of life." In short, had this intellectual climate continued, humanity's future would have been little more than a clone of its past.

Fortunately, these circumstances were not destined to last. By the seventeenth and eighteenth centuries, Western society was undergoing radical and favorable changes. And the changes that occurred gave rise to the notion that progress was a real force urging civilization forward. Surpluses of consumer goods were replacing substandard living conditions in most of Western Europe. Rigid class systems were giving way to social mobility. In nations like France and the United States, the lofty principles of democracy and popular sovereignty were being painted in broad, gilded strokes over the fading canvasses of monarchy and despotism.

But more significant than these social, economic, and political changes, the new age witnessed a rebirth of science. Centuries of scientific stagnation began crumbling before a spirit of scientific inquiry that spawned undreamed of technological advances. And it was the discoveries and inventions of scores of men and women that fueled these new technologies, dramatically increasing the ability of humankind to control nature—and, many believed, eventually to guide it.

It is a truism of science and technology that the results derived from observation and experimentation are not finalities. They are part of a process. Each discovery is but one piece in a continuum bridging past and present and heralding an extraordinary future. The heroic age of the Scientific Revolution was simply a start. It laid a foundation upon which succeeding generations of imaginative thinkers could build. It kindled the belief that progress is possible as long as there were gifted men and women who would respond to society's needs. When Antonie van Leeuwenhoek observed *Animalcules* (little animals) through his high-powered microscope in 1683, the discovery did not end there. Others followed who would call these "little animals" bacteria and, in time, recognize their role in the process of health and disease. Robert Koch, a German bacteriologist and winner of the Nobel Prize in Physiology and Medicine, was one of these men. Koch firmly established that bacteria are responsible for causing infectious diseases. He identified, among others, the causative organisms of anthrax and tuberculosis. Alexander Fleming, another Nobel Laureate, progressed still further in the quest to understand and control bacteria. In 1928, Fleming discovered penicillin, the antibiotic wonder drug. Penicillin, and the generations of antibiotics that succeeded it, have done more to prevent premature death than any other discovery in the history of humankind. And as civilization hastens toward the twenty-first century, most agree that the conquest of van Leeuwenhoek's "little animals" will continue.

The *Encyclopedia of Discovery and Invention* examines those discoveries and inventions that have had a sweeping impact on life and thought in the modern world. Each book explores the ideas that led to the invention or discovery, and, more importantly, how the world changed and continues to change because of it. The series also highlights the people behind the achievements—the unique men and women whose singular genius and rich imagination have altered the lives of everyone. Enhanced by photographs and clearly explained technical drawings, these books are comprehensive examinations of the building blocks of human progress.

LASERS

Humanity's Magic Light

LASERS

Introduction

One day in 1960, something happened that seemed almost magical to many people. Dr. Theodore Maiman put into operation a device that gave off a thin, bright red beam of light. This remarkable device was the world's first laser. It was called a ruby laser because Maiman passed ordinary light through a ruby to produce the laser light. Since that day, hundreds of different kinds of lasers have been made, and thousands of practical uses have been found for these modern "supertools."

Until the invention of the laser, using powerful beams of light was an idea that appeared mainly in science fiction. The first important example was H.G. Wells's classic *The War of the Worlds* (published in 1898). In the novel, sinister martians used a terrifying heat ray to attack the earth.

Later, in the early comic strips, space heroes like Flash Gordon and Buck Rogers used deadly ray guns to fight their archenemies. Certainly, most people today are familiar with the exploits of Captain Kirk and the Starship *Enterprise*. Fans of the "Star Trek" TV show and movies know all about phasers and photon torpedoes, those fabulous twenty-third century weapons of light. Equally famous are the light swords wielded by Luke Skywalker and Darth Vader in the popular *Star Wars* films.

...TIMELINE: LASERS

1 ■ 1880
Alexander Graham Bell builds the photophone.

2 ■ 1905
Albert Einstein proposes that light is made up of both particles and waves.

3 ■ 1917
Einstein describes the process of stimulated emission.

4 ■ 1934
Norman R. French invents the light pipe.

5 ■ 1947
Charles Townes begins microwave experiments at Columbia University.

6 ■ 1954
Townes builds the first maser.

7 ■ 1957
First designs are drawn for the laser.

8 ■ 1960
Theodore Maiman builds the first laser.

Of course, all the devices mentioned are destructive. This can be misleading, for science fiction is filled with examples of light being used for constructive purposes as well. For instance, "Star Trek" characters use a small light source to perform operations and use beams of light to take apart thick metal walls or repair broken circuits.

Laser light is very different from ordinary natural light. The laser produces light that has been amplified, that is, made brighter and more powerful. It took human beings, in their never-ending search for new knowledge, to invent the laser and take advantage of its marvelous properties. Just as Captain Kirk and the *Enterprise* go "where no man has gone before," scientists saw a light no one had seen before and put it to use for the good of all people.

On that day in 1960, when the ruby laser began to glow, a new era began for humanity. The light no one had seen suddenly changed science fiction into science fact. As lasers continue to advance and do more things, it seems more likely that the world of Captain Kirk may be just around the corner.

9 ■ 1970
First long-distance fiber-optic system is built.

10 ■ 1974
First tunable, continuously operating laser is built.

11 ■ 1981
Laser beacon to prevent mid-air collisions is developed.

12 ■ 1985
Laser tracking system is tested by space shuttle *Discovery*.

13 ■ 1986
Young Sherlock Holmes becomes the first motion picture to use laser beams to print images directly onto film.

14 ■ 1988
After long legal battle, Gordon Gould receives major laser patent applied for thirty years earlier.

Pre-Laser Ideas and Inventions

A laser is a device that produces an unusually powerful beam of light that does not exist on its own in nature. Today, this light is used to perform thousands of useful tasks.

A laser can give off a light beam of awesome power. It can blast through a thick metal wall or bore a hole in a diamond, the hardest substance known. But a laser can also be easily controlled. Some lasers can measure things seen only under a microscope or even perform delicate eye operations.

Every day, lasers are used in communications, in industry, in hospitals, and in the field of entertainment. In fact, lasers make our lives better and easier in so many ways that scientists have come to refer to these devices as the supertools of the modern age.

The invention of the laser was the result of many ideas and discoveries, each building upon the ones that came before it. These ideas and discoveries go back more than one hundred years. Scientists did not purposely set out to invent the laser; in fact, no one seriously considered such a device until just a few years before the first laser was built.

Most of the ideas that led to the invention of the laser were the results of attempts by scientists to learn more about

Special effects technicians often use laser light to thrill audiences. The entertainment industry depends on laser projections like these for movies and concerts.

light and how it behaves. As time went on, the idea that light might be made more powerful, or amplified, became more important to scientists. Finally, a few researchers managed to put all these accumulated ideas together in a very special way. Only then did it occur to scientists that something like a laser could be built.

The Invention of the Photophone

The first important attempt to get light to perform a task that lasers perform today came in the year 1880. Alexander Graham Bell performed an experiment that showed how light might be used to carry a person's voice from one location to another. To accomplish this, Bell used a device he called a photophone, which consisted of a thin mirror, a receiver that could detect light, some wires, and an earpiece. Bell placed the mirror so that sunlight reflected off of its surface and traveled more than one hundred feet to the receiver. When a person spoke near the delicate mirror, it jiggled, or vibrated. This caused the sunlight being reflected into the receiver to vibrate also. The receiver then changed the light vibrations into an electrical signal, which traveled through the wires to the earpiece.

Unfortunately, the photophone did not work very well. Bell's receiver was very crude compared to the ones used today. Also, the device relied on sunlight, which varies in brightness from hour to hour and from day to day. Obviously, on cloudy days or at night, the device could not be used at all.

Bell also had to contend with the fact that scientists at that time still did not know enough about light in order to use its power. What they did know was that

Alexander Graham Bell tried to harness the power of light when he invented the photophone in 1880. The photophone relied on mirrors and sunlight to transmit sound.

light always travels at a set speed, which happens to be about 186,000 miles (300,000 kilometers) per second. This is so fast that a beam of light can race around the earth almost seven-and-a-half times in a second. But in order to use light as a tool, it was not enough to know how fast light travels. People also needed to know what light is made of.

In the 1600s, the English scientist Sir Isaac Newton suggested that light was made up of tiny particles. This explanation was the particle theory of light. At about the same time, the Dutch scientist Christian Huygens said that light might be composed of waves, similar to the ocean waves that roll onto a beach. This was the wave theory of light.

Unfortunately, in Bell's day, scientists still argued over which of these theories was correct. Through a mixture of hard

claimed that the particles, called photons (from the Greek word for "light"), moved along in wavelike patterns. Later, other scientists performed experiments that proved Einstein was right. Einstein himself then went on to predict some more startling things about light. These predictions were about how photons are made and became the next step in the journey toward the invention of the laser.

Einstein agreed with some other scientists of his day about how light sources (like candles, light bulbs, or the sun) produce photons. The researchers thought that atoms give off photons. Atoms are the tiny building blocks that make up rocks, trees, people, and everything else in the known universe. Some form of energy such as heat, electricity, or chemical energy, might excite an atom, or make it more energetic. It would then emit, or give off, a photon. Afterward, the

In the seventeenth century, Sir Isaac Newton proposed the theory that light is composed of many tiny particles.

work and genius, however, Bell managed to build a working photophone. The device itself operated on principles different from those that would later be used in the laser. Yet Bell had demonstrated that light might someday be used in communications. This idea prepared the way for the next step in the trip to the laser.

Einstein and Stimulated Light

In 1905, Albert Einstein, a German researcher, announced his own theory about light. He said it was made up of both particles and waves. Einstein

Albert Einstein developed a new theory of light in 1905. This theory states that particles of light, called photons, move in waves.

HOW PHOTONS CAUSE LIGHT

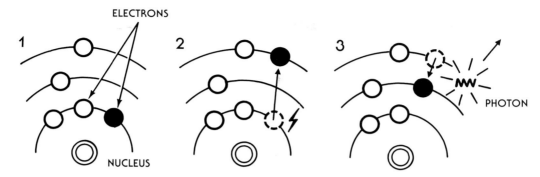

Atoms give off light when they are "excited" by an outside energy source, such as heat or light. A beam of light is a form of electromagnetic energy. Electromagnetic energy is composed of billions of photons, or tiny bursts of energy, that travel in waves. The diagram above shows how atoms emit photons.

1.   Every atom contains electrons that revolve around its nucleus in various orbits, or energy levels. The level closest to the nucleus is the lowest en-ergy level, and the one farthest from the nucleus is the highest.

2.   When an atom is "excited" by an external energy source, at least one of its electrons picks up extra energy and jumps to the highest energy level.

3.   An electron that has jumped to the highest level will soon return to a lower level. When it does, it rids itself of the extra energy. That energy is emitted as a photon.

atom would go back to its normal, unexcited state. Because there are huge numbers of atoms, they give off equally large numbers of photons. A one hundred-watt light bulb gives off about ten trillion photons every second.

In 1917, Einstein suggested that while an atom was excited, it might be coaxed, or stimulated, to produce a photon. If enough atoms could be excited and stimulated, a great number of photons might be produced. A beam of light made up of so many photons would be highly concentrated and, therefore, brighter and more powerful. Einstein called this process stimulated emission of light. Because it can produce light that is boosted in power, or amplified, stimu-lated emission is the basic principle of laser operation.

Scientists had the basic information they needed to build a laserlike device as early as 1917. But no one actually attempted to build such a device for two reasons. First, researchers thought it would be too difficult and expensive, and they were right. The advanced machinery they needed did not exist at the time and would have to be developed, piece by piece. In particular, a way would have to be found to physically stimulate the atoms to produce photons. Also, the process of exciting and stimulating the atoms would have to be kept going somehow. No one had any idea how these things might be done. The necessary

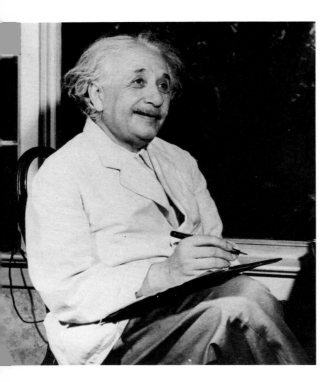

Einstein described how atoms could be stimulated to produce a bright and powerful beam of light. Using this knowledge, other scientists invented the laser.

research would take many years and cost a lot of money. Of course, such time and money would be worthwhile if the idea promised to be useful. This brings us to the second reason no one built a laser any earlier, a reason that might seem surprising. Scientists did not think that the idea of amplifying light would lead to anything practical or useful. They did not foresee the laser. As it turned out, lasers were developed in a more roundabout way—as a result of radar.

Experiments with Microwaves

During World War II (1939-1945), scientists worked hard at improving radar, a

tracking device that had been invented a few years earlier. Radar sends out a beam of microwaves that bounces off nearby objects. Microwaves are similar to visible light. Both microwaves and light are types of radiation. They both travel at 186,000 miles per second, and both are made up of particles that move in waves. But, unlike light, which we can see, microwaves are invisible to us.

After a radar beam bounces off an object, the reflected microwaves return and register on a screen. By studying the screen, a radar operator can tell the general size and distance of the object. This is how soldiers during World War II located enemy airplanes.

As the war dragged on, enemy scientists learned to "jam," or interfere with, American radar beams. American military scientists then tried to find ways of producing more powerful microwaves. Researchers believed that these more energetic waves would be able to blast through the enemy jamming.

The American scientists, led by Dr. Charles Townes, performed several experiments with the more powerful microwaves. Unfortunately, Townes and the others found that this type of microwave radiation does not work very well for radar. The waves are too easily absorbed by water vapor in the air. This means that as the beam travels farther, more of it gets absorbed. Eventually, the beam gets too weak to do any good. But, in a way, these experiments had not failed. They caused Townes to become very interested in microwaves. Townes did not realize it at the time, but he was now on the trail that would eventually lead to the laser.

In 1947, Townes began teaching at Columbia University in New York City. There, in the university's labs, he continued with his microwave experiments. He remembered what Einstein had said about the stimulated emission of visible light: that many atoms could be stimulated to produce many particles of light. Since microwaves are so similar to light, Townes reasoned that stimulation might also produce many microwave particles. If the microwaves could be produced by stimulation, perhaps enough of them could be built up to get an amplified beam. A microwave-amplifying device might be built to accomplish this task.

But what would be the purpose of such a device? Townes was not sure it would have any practical uses. But he knew it would aid scientists in studying how atoms give off radiation. In short,

the device would be useful as a research tool. The work continued, but no real breakthroughs came for a couple of years.

The Maser Sets the Stage for the Laser

Then, in 1951, while sitting on a park bench, Townes had a brilliant idea. He realized it might be possible to use molecules of ammonia to produce a powerful microwave beam. A molecule consists of two or more atoms that are connected together.

Townes figured that when molecules of ammonia became excited (by heat,

Charles Townes and James P. Gordon show off a maser device. In the maser, heat excites ammonia molecules and causes them to emit a concentrated beam of microwaves.

HOW A MASER WORKS

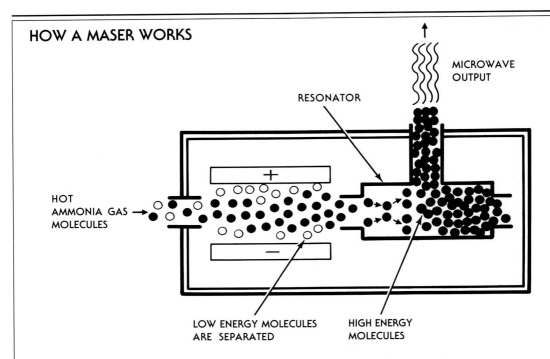

RESONATOR

MICROWAVE OUTPUT

HOT AMMONIA GAS MOLECULES →

+

−

LOW ENERGY MOLECULES ARE SEPARATED

HIGH ENERGY MOLECULES

A maser amplifies, or increases, the number of photons that cause invisible electromagnetic waves known as microwaves. First, heated ammonia gas is pumped into the maser. There, "unexcited," or low-energy, molecules are drawn to the sides of the maser. Only "excited," or high-energy, ammo-nia molecules flow into the resonator, or vibrating chamber. In this chamber, the "excited" molecules begin to emit micro-wave photons. These photons bounce around in the resonator, striking the am-monia molecules so that they remain "excited" and produce more and more microwaves.

electricity, or chemical energy), they could be stimulated to emit microwaves of the type he was working with. He knew this process would be almost identical to the one Einstein described for stimulat-ing visible light. The only difference was that Townes would be using microwaves instead of light. He calculated that if the ammonia molecules could be kept in an excited state long enough, they might be stimulated to produce more and more microwaves. Eventually, the waves would become concentrated and more power-ful. In short, the microwaves would be amplified.

Townes decided to go ahead and try building a working model. He enlisted the aid of two other researchers, Herbert J. Zeiger and James P. Gordon. The three men encountered several technical prob-lems, but after three years of hard work, they managed to solve them.

In 1954, Townes and the others finally had a working device that operated in the following way: The men heated some ammonia gas until many of the molecules became excited. Then, they separated the excited molecules from the unexcited molecules. Next, the scientists made the excited molecules flow into a chamber called the resonant cavity (or resonator).

It was inside the resonator that the stimulation of the molecules took place. As the excited ammonia molecules began to emit microwave particles, the particles began to bounce back and forth inside the chamber. When one of these particles came near an excited molecule, the molecule suddenly gave off its own particle. Thus, the particles themselves stimulated the production of more particles. Soon, the number of particles doubled, then doubled again and again, until the microwaves in the chamber had become very powerful.

The entire process described here took only a tiny fraction of a second. Out of a hole in the resonator shot a strongly amplified beam of microwaves. The scientists called their invention a maser. The letters of this acronym stand for *M*icrowave *A*mplification by *S*timulated *E*mission of *R*adiation.

Although Townes and the others were the first to actually build a maser, scientists in the Soviet Union had also been working on such a project at about the same time. However, the Soviets could not manage to build a working model before the Americans finished their own.

As Townes had suspected, the ammonia maser did not have many practical uses. But it was a research tool scientists could use to study microwaves. In addition, the device did do two things very well. First, the scientists found that the ammonia molecules in the maser vibrated at a steady, unchanging rate. Therefore, the maser could be used as a reliable timekeeping device. Second, because the maser was an amplifier, it could boost the weak microwave signals given off by distant stars. This made it easier for astronomers to study such signals and learn more about stars.

Townes and other researchers did not know it then, but the most important thing about the maser was that it set the stage for the development of the laser. All the right ingredients needed for the laser had been combined in the maser, except for the most important one—light. The scientists would now take the concepts of the maser and amplified microwaves and use them to produce the laser and amplified light. This would turn out to be a very different kind of light, one that had never been seen before. The modest little maser was about to give birth to a device with thousands of uses, a tool with the power to change the world.

The Laser

In the mid-1950s, maser research grew and scientists built more efficient versions of the device. These improved masers helped the scientists understand the way atoms behave and, of course, increased their knowledge about microwaves. But microwaves are only one type of radiation. The success of the maser made scientists want to experiment with other types of radiation, including light. This experimentation with light would eventually lead to the invention of the laser.

Soon after the introduction of the maser, some researchers suggested taking the maser idea a step further. They began talking about a device that would stimulate atoms to emit photons of light, exactly as Einstein had described. The photons would then be amplified to produce a powerful beam of light.

No one at the time knew exactly what such a device would be good for. Most researchers assumed that, like the maser, it would aid in the study of astronomy. Some suggested possible uses in communications and a few other areas. But these ideas were vague and rather limited in scope. As it turned out, scientists did not realize just how useful lasers would be until after the first ones had been built.

The Laser's Many Inventors

One researcher who wanted to learn how to amplify light was the father of the maser, Charles Townes. In September of 1957, Townes drew a design for a device

he called an optical maser (the term *laser* had not yet been coined). He then called another scientist, Arthur Schawlow (who happened to be his brother-in-law). The two men began drawing up more detailed plans for the optical maser.

But Townes and Schawlow were not the only ones working on the laser idea. In the Soviet Union, Nikolai Basov and Aleksander Prokhorov also drew designs for such a device. In addition, a graduate student at Columbia University named Gordon Gould thought about developing his own light-amplifying device. Commenting later, Gould said, "I was immediately electrified by the idea and knew it was going to be important."

In November of 1957, only two months after Townes had sketched the optical maser design, Gould wrote down in notebooks all his ideas for his proposed invention. The first thing he wrote was a name for the device. He called it a laser, which stood for *Light Amplification by Stimulated Emission of Radiation*. Other researchers thought of this name on their own, but apparently Gould was the first to coin the term.

In his notebooks, Gould made some unusual predictions about the future use of lasers. It seems that he was the only person at the time to realize the enormous power potential of the device. For instance, he wrote about laser beams reaching temperatures hotter than the surface of the sun. Such beams, he suggested, might someday be used to trigger nuclear fusion, the process in which atoms

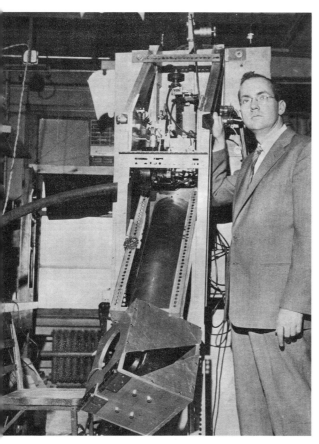

Charles Townes stands by the first maser used for radio astronomy at Columbia University in New York. Success with the maser inspired Townes to experiment with light radiation.

lawyer who specialized in patents. When the government grants a person a patent for an invention, it recognizes that person as the original inventor. But there is more involved than just being first. The person who holds the patent can make a lot of money if the invention is successful.

Unfortunately for Gould, the lawyer did not understand the information in the notebooks. The man gave Gould the mistaken idea that he needed a working model of his invention in order to get a patent. Gould was not sure what to do next, so he just waited. This turned out to be a mistake because in the meantime Townes and Schawlow had been hard at work. They applied for their own patent in the summer of 1958. They also wrote a paper explaining their ideas and had it published in a famous science magazine, an accepted practice in science. This tells other scientists, as well as the public, who was first with the idea.

Gould had not published a paper. He finally did apply for a patent, but it was almost a year after Townes and Schawlow had applied for theirs. Therefore, no one believed Gould later when he claimed he had come up with the idea for lasers on his own.

are combined to produce enormous amounts of energy. Today, scientists in many labs around the world are working to make this happen.

Gould Encounters Problems

Hearing that Townes might also be working on lasers, Gould got worried. Naturally, he did not want someone else to get credit for what he considered to be his own invention. He showed his notes to a

The First Laser Is Built

By 1960, many scientists, including Townes and Schawlow, the two Soviets, and Gould, had asked for laser patents. In addition, the paper published by Townes and Schawlow had caused widespread interest in lasers in the American scientific community. Researchers in labs around the country raced to be first to construct a working model. But despite this flurry of activity, no one seemed very close to actually building a laser.

Charles Townes poses with a ruby laser. Unlike the maser, which uses heat to excite ammonia molecules, this laser depends on light to excite the molecules of the ruby.

which excites the atoms or molecules, had been heat. But heating a ruby would only cause it to break. Then, Maiman noticed that light passed through a ruby. Perhaps ordinary light itself could be used as the excitation device. Maiman rigged a powerful flash lamp so that it ran through a glass tube. The tube was bent into the shape of a coil and wound several times around the ruby. When the lamplight flashed, the photons excited the atoms in the ruby. These excited atoms became stimulated and gave off their own photons.

Maiman had accomplished the first step in the lasing process. He had stimulated the atoms in the medium to produce photons. But he needed to greatly increase the number of photons being

As a result, the announcement of the operation of the first laser on July 7, 1960, took the scientific world completely by surprise. The device had been built by an unknown researcher who had worked totally on his own. He was Theodore H. Maiman, a researcher at the Hughes Aircraft Company in Malibu, California.

Maiman's laser was small (only a few inches long) and not very complicated. The core of the device consisted of an artificial ruby about one-and-a-half inches long, so Maiman called his invention the ruby laser. The ruby acted as the lasing medium, which is the substance that supplies the atoms or molecules to be stimulated. (In Townes's maser, the medium had been ammonia gas.)

Maiman knew that the atoms inside the ruby would need to be excited somehow. In the maser, the excitation device,

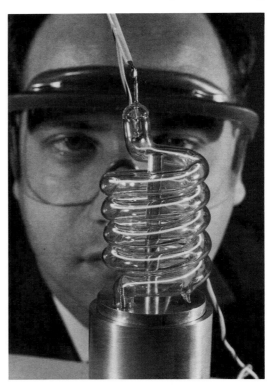

Theodore Maiman of Hughes Aircraft Company stands behind his ruby laser. A core of synthetic ruby is surrounded by a spiraling flash lamp.

THE RUBY LASER

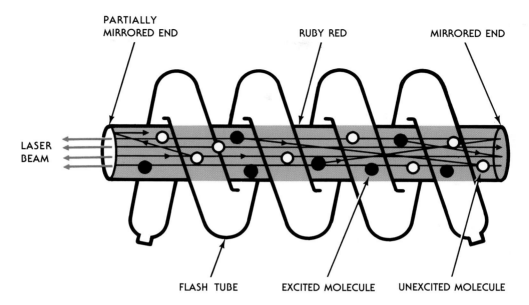

PARTIALLY MIRRORED END

RUBY RED

MIRRORED END

LASER BEAM

FLASH TUBE EXCITED MOLECULE UNEXCITED MOLECULE

A ruby laser works much like a maser. But instead of microwaves, it produces an intense beam of visible light. This occurs when the molecules of an artificial ruby are "excited" by a flash of ordinary light from a flash tube, which surrounds the ruby. The "excited" ruby molecules begin to emit photons of visible light. The photons reflect off of mirrors located at both ends of the ruby. The reflected photons continue to excite more and more ruby molecules, thus producing more and more photons. This production of photons amplifies the light inside the ruby. The mirror on one end of the ruby is only a partial, or half-silvered mirror, so some of the light passes through it. This light is the laser beam, or beam of amplified light.

produced. In other words, he needed to amplify the light. In the maser, amplification had occurred inside the resonant cavity where the particles bounced back and forth. In Maiman's laser, the ruby itself acted as the resonator because it contained the excited atoms.

But there was nothing in the ruby for the photons to bounce off of. Furthermore, a ruby is nearly transparent. The photons created would just pass through and escape. Maiman had to figure out how to make the photons bounce back and forth and also how to keep them from escaping. He accomplished both of these goals in a very simple way. The ruby was shaped like a cylinder, something like a long soup can (but much smaller). Maiman coated each end of the cylinder with silver so that the ends became mirrors pointing toward the center of the ruby. Because the mirrors faced each other, the photons now bounced back and forth through the ruby over and over again.

The amplification process in Maiman's laser was almost identical to that of the maser. The photons kept passing near excited atoms and stimulating them to produce more photons. Eventually, enough photons had been produced to form a beam of light. But how could the beam get by the mirrors and escape? Fortunately, Maiman had already solved this problem when he coated the cylinder ends with silver. He thinly coated one end so that it became only a partial mirror. It reflected some photons back into the ruby at the same time it allowed others to escape. The ones that escaped became the actual laser beam. To Maiman's delight, the beam of strange, deep red light shot out of the laser and registered on a nearby detector. As in the maser, the entire process happened extremely fast. In fact, it took only a few millionths of a second.

Birth of the Laser Era

Of course, Maiman immediately went to a scientific journal to publish the results of his experiment. But the editors of the journal did not realize the importance of the work and refused to publish the paper. Maiman then went to the editors of a British journal called *Nature*. They decided to publish his article, which was only about three hundred words long. Those three hundred words excited scientists all over the world. Dozens of labs began to build their own devices. The era of lasers had begun.

Scientists quickly realized that other materials besides ruby could be used as laser mediums. The first gas laser used a

A scientist experiments with a water vapor laser. After the ruby laser was invented, researchers discovered that a variety of other materials, such as crystals and gases, could be used as laser mediums.

The helium-neon laser was the first gas laser to be developed after Theodore Maiman published the results of his ruby laser research.

mixture of helium and neon gases. It was built only a few months after Maiman's paper appeared in the journal. Soon, researchers constructed lasers with many other types of mediums: crystals, carbon dioxide gas, vaporized metal, and even colored dyes.

As more and different types of lasers were developed, scientists found more ways to use them. Soon, many people started thinking of tasks for lasers to perform. What had once seemed like a useless idea now appeared to have almost unlimited uses. People in communications, industry, medical labs, and the military all rushed to find ways they might benefit from the new type of light. Astronomers wanted to use lasers to study the sun and stars. Engineers wanted lasers to cut and weld metal parts in factories. Doctors saw potential for lasers in

performing eye operations and burning away tumors. Military leaders thought lasers might be developed into death rays to shoot down enemy planes and satellites. There seemed to be hundreds of possible uses for the new tool.

A Question of Money

Obviously, all this new laser research cost a great deal of money. But government officials as well as business leaders felt the valuable invention would prove to be worth every penny they spent on it.

The invention of the laser also promised to be valuable to the inventor. As is true with other patent holders, the person recognized as the inventor of the laser would have a chance to make a lot of money. Each time an invention is built and operated by someone else, the inventor receives a payment called a royalty. As a rule, the person recognized as the inventor of a device is the one granted the major patent or patents. When a single person or group applies for a patent, recognizing the inventor(s) is easy. But, in the case of the laser, several researchers had come up with the idea at about the same time. This made it difficult for the U.S. Patent Office to decide who the actual inventor was.

As it turned out, Townes and Schawlow received patents for the basic laser principle, and Maiman received a patent for the ruby laser. All three men had been recognized as inventors of the device and could collect royalties. However, because he had applied so late, Gould did not receive a patent and could not collect royalties. Upset at being left out, Gould took the case to court but the legal battle dragged on for years.

A Hughes Aircraft laser technician uses a ruby laser to pierce a hole through a sheet of hard tantalum metal.

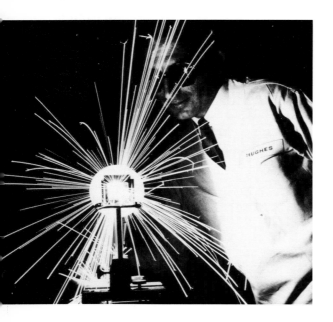

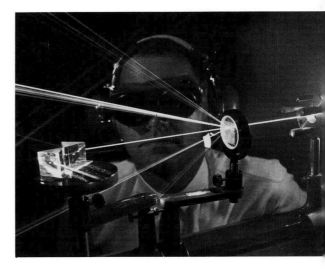

The argon laser emits bright and narrow beams of light that are useful for processing computer information and for carrying voice and television signals.

Unfortunately, Gould had even more to be upset about. In 1964, the famous Nobel Prize for physics was awarded to the men who had done the original designs for lasers. Three men shared the prize: Townes and the two Soviets, Basov and Prokhorov. Again, Gould had been left out.

Gould refused to give up. He kept insisting that he too had invented the laser. He went into debt when he had to borrow large sums of money to continue the legal fight. But the time, effort, and money eventually paid off. In 1977 and 1979, he received patents for two small parts of the lasing process. Finally, in 1988, the Patent Office granted him the major patent he had applied for back in 1959, allowing him to receive royalty checks from such giant companies as AT&T, General Motors, and 3M.

And so, along with Townes, Schawlow, Maiman, Basov, and Prokhorov, Gould is at last recognized as one of the founding fathers of the laser.

The Laser and the Military

The U.S. military became interested in lasers even before the first lasers had actually been built. When military leaders heard that the new device might produce very hot beams of light, they immediately dreamed of developing beam weapons. They hoped these weapons would do many things that ray guns had done in science fiction stories. Perhaps such weapons might blast holes through enemy soldiers and tanks or even shoot down planes and satellites.

Much of the military's early exposure to the laser idea came from Gordon Gould. Even though he was having trouble getting a patent, he continued to work on lasers. When he left Columbia University in 1959, he went to work for Technical Research Group, a company in Syosset, N.Y., that did research on radar, missiles, and other military projects. Gould got TRG very interested in lasers. TRG then asked the Department of Defense for $300,000 to do research on laser weapons. The military was so fascinated by the idea of beam weapons that it gave the company almost a million dollars, more than three times the amount that had been requested.

A few months later, when Maiman actually built the ruby laser, his own company, Hughes Aircraft, began working on laser weapons. Soon, the various branches of the military—army, navy, and air force—gave money to several other companies to develop such weapons. Between 1962 and 1968, the army alone spent almost $9 million on laser research.

But could such death rays actually be built? The answer, at least at that time, was no. Scientists found that lasers did indeed produce very concentrated beams of light. But building the kinds of weapons the military wanted turned out to be much more difficult than everyone had figured.

Ray Guns

In the case of hand-held ray guns, too much energy would be required for devices so small. To produce enough power, such guns would have to be so huge no one could carry them. Also, the guns themselves would get very hot, hot enough to badly burn the hands of the people holding them. Researchers showed that, for the time and money required to build them, ordinary rifles would be more effective hand-held weapons than lasers.

Lasers that could shoot down planes or satellites seemed more promising. These would also need to be quite large, but it did not matter since they would not have to be hand-held. In the the 1960s, however, the lasing mediums being used did not produce beams powerful enough to shoot down enemy aircraft. Later, researchers tried other lasing mediums, such as a mixture of the elements fluorine and hydrogen. This produced a lot of power but had some serious problems. The mixture explodes easily and without warning. Also, the exhaust gas is hard to get rid of and kills anyone who breathes

The hydrogen-fluoride laser was a powerful but dangerous device. Scientists discovered that it exploded easily and without warning.

it. The scientists encountered many other such problems.

But military leaders continued to pour money into laser weapons research. They knew that such weapons would have some clear advantages over normal bullets and missiles. In the first place, when firing a bullet at a moving target, one has to aim a bit ahead of the target. This is because the target itself moves ahead while the bullet is racing toward it. Since gravity pulls the bullet downward, one also has to aim a bit above the target. In the middle of a battle, with all the smoke, noise, and confusion, hitting a moving target can be a difficult task.

But a laser beam moves at the speed of light. This means the beam can travel a mile in only six-millionths of a second. Even if an airplane was traveling at the speed of sound, it would only move about one-sixteenth of an inch, while the laser beam covered the mile. Military experts realized that more hits could be scored by laser weapons than by ordinary guns and missiles.

There seemed to be other benefits of beam weapons. They could be operated with very little fuel. And, the beam could be bounced off mirrors, so only the mirrors needed to be moved when switching to a new target (instead of moving the whole weapon). Also, a laser beam stays concentrated over long distances, so it might be able to hit targets hundreds of miles away.

In 1973, the army finally succeeded in shooting down a drone (a remote-controlled plane). But military leaders did not consider the test a complete success. The drone did not move as fast as enemy

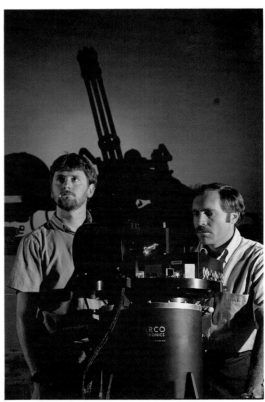

The military uses laser tracking systems such as this one to locate targets and measure distances to them.

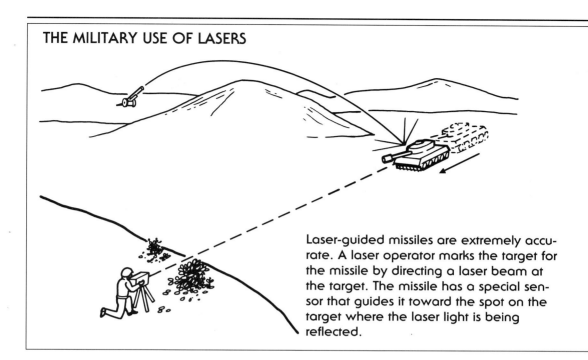

THE MILITARY USE OF LASERS

Laser-guided missiles are extremely accurate. A laser operator marks the target for the missile by directing a laser beam at the target. The missile has a special sensor that guides it toward the spot on the target where the laser light is being reflected.

aircraft. Also, the program leading up to the test had been very expensive. In fact, laser weapons in general seemed to be almost too expensive. At that time, a modern tank cost about $1 million to make; experts estimated that a battlefield laser would cost nearly $10 million. But the military still pressed on with laser experiments.

By 1980, the U.S. government had spent hundreds of millions of dollars on laser research. Unfortunately, there had been only a few practical results. Looking back, many experts feel the military moved too quickly on lasers. It did not allow enough time for developing the new technology before it demanded results.

The Real Battlefield Lasers

The laser research that did bring results for the military did not produce beam weapons. Instead, scientists came up with devices to improve the accuracy of normal, conventional weapons. Of these devices, laser range finders and bomb designators found the most widespread use.

The range finder turned out to be the army's greatest laser success. The finder calculates the distance, or range, to a desired target. It does this by measuring how long a small burst of laser light takes to travel to the target. This practical tool can be either hand-held or mounted on a tank. Obviously, if a soldier knows the exact distance to his target, he has a much better chance of hitting it. By the mid-1970s, Hughes Aircraft was building more than $50 million worth of laser range finders for the army each year.

The laser bomb designator works by shining a low-power laser beam at the desired target. A bomb is released, either from an airplane or from a ground-based missile. This is known as a "smart" bomb because it carries a sensor that can detect a laser beam. The smart bomb homes in on the beam and destroys the target. The army first used the laser designator on the battlefield in 1972 during the Vietnam War.

Lasers have also proved to be successful in battle simulations. These are mock, or pretend, battles that are staged to give soldiers practice for the real thing. Before lasers, these simulations had not been as realistic as military officials would have liked. Obviously, the soldiers could not really fire at each other. Referees had to decide who had or had not been hit by the opposing team.

In laser simulations, soldiers fire special guns at each other. These guns shoot bursts of light. Sensors are attached to each soldier who fights in the battle. Sensors are also attached to tanks, trucks, or any other vehicles used in the mock fighting.

When a burst of light is fired and hits a sensor on an "enemy" soldier, the sensor registers the light and everyone knows immediately that the soldier is "dead." They know when a tank or truck has been destroyed because the sensors mounted on vehicles give off a cloud of smoke when hit. Many companies now build such battle simulators, which are used by armies all over the world.

The Submarine Laser

One dramatic use for the military laser is in the area of submarine communications. Submarines often patrol in

Nova, the world's most powerful laser, is the latest in a series of increasingly powerful lasers developed by Lawrence Livermore Laboratories. The laser is used in experiments with fusion reactions.

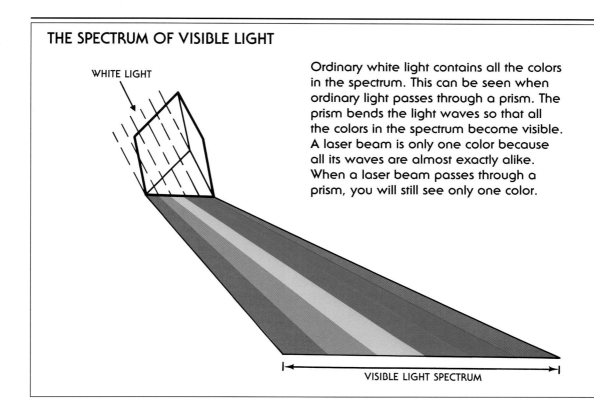

THE SPECTRUM OF VISIBLE LIGHT

WHITE LIGHT

Ordinary white light contains all the colors in the spectrum. This can be seen when ordinary light passes through a prism. The prism bends the light waves so that all the colors in the spectrum become visible. A laser beam is only one color because all its waves are almost exactly alike. When a laser beam passes through a prism, you will still see only one color.

VISIBLE LIGHT SPECTRUM

enemy waters. In the past, the only way an admiral could get a message to a sub was by using ordinary radio. But this has two serious disadvantages. First, radio waves do not travel well underwater and require large antennas to broadcast them long distances. Second, there is always the risk that the enemy will pick up the signal. If this happens, the sub's location is immediately revealed.

In this case, a laser is more effective than radio because of a unique and important quality of laser light: it is monochromatic. This means that it shines in only one color. It is very different from ordinary white light, which is made up of all the colors bunched together.

To send a message to the submarine, the navy uses a laser that gives off a monochromatic beam of blue-green light. This particular shade of blue-green

travels easily through ocean water. The beam carrying the message is transmitted to a satellite orbiting high above the ocean. The satellite relays the beam down to the sub, which has a special receiver that registers only blue-green light. In less than a second, the sub's computer decodes the signal so the crew can read the message.

If there are any enemy lookouts nearby, it is unlikely they will know about the signal beam. In the first place, the satellite flashes the beam for only a few millionths of a second. This is not enough time for the lookouts to see it. Even if the lookouts have a receiver that detects laser light, it has to be tuned to receive the exact shade of blue-green in the beam. The receiver also has to be underwater near the sub because that is where the satellite aims the beam. Meeting these

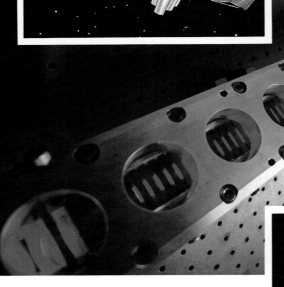

Left: An artist depicts a Star Wars battle in space between the superpowers. A laser from the satellite shoots an enemy missile.

Below: This free-electron laser is used in Star Wars weapons.

Right: A Star Wars laser satellite locates and shoots down an incoming missile.

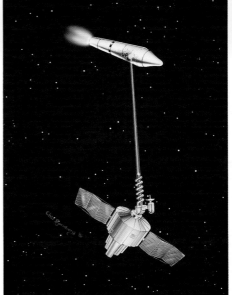

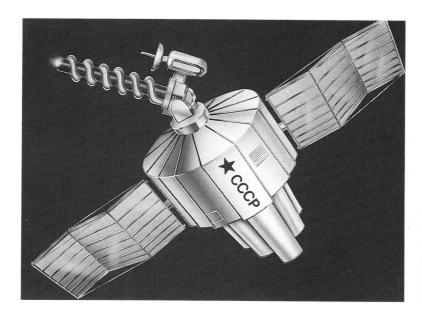

An artist's drawing shows a Soviet Union laser satellite used to attack American weapons that might be launched into space.

conditions would be very difficult for the lookouts, so chances are good the sub will get the message and still remain undetected. The navy feels that this form of laser communication is very promising and continues to study ways to improve it.

The Future of Laser Weapons

Meanwhile, the Department of Defense has not given up on developing beam weapons. In fact, research in satellite and rocket lasers increased dramatically in the 1980s. This was mainly due to President Ronald Reagan's call for the building of a defense system nicknamed Star Wars. Reagan's idea was to stop enemy nuclear missiles from hitting the United States during a nuclear war. He felt this could be done by destroying the missiles while they are still high above the earth's atmosphere. Laser sensors mounted on satellites would detect incoming missiles and plot their paths. The missiles would

then be shot out of the sky, either by beam weapons or by "smart" American missiles that home in on designator beams.

The Star Wars defense system has been the subject of debate among scientists, politicians, military personnel, and the American public. Some scientists believe the long-term benefits of such a program are questionable and that the expense involved in implementing it is too high. Others believe that Star Wars is necessary to U.S. military strategy and is well worth the money. Whatever the future of Star Wars, the debate is likely to continue for many years.

There is one thing that everyone can be sure of. Lasers have found a permanent home in the arsenals of the world's armies. Research into the military use of lasers will continue, and the wars of the future, if any, will be fought very differently than the wars of the past. It is likely that someday, several different kinds of lasers will have become standard equipment for every soldier, sailor, and pilot.

CHAPTER 4

The Laser Toolbox

In 1969, the *Apollo 11* astronauts became the first men ever to walk on the moon. Before blasting off on their return flight, they left behind a bizarre-looking mirror. A short time later, scientists on earth claimed that the strange mirror had revealed to them the distance from the earth to the moon. The figure the mirror gave them was accurate to within the length of a person's finger.

At first glance, the moon mirror may seem mysterious, perhaps even magical. But there was really no magic involved. National Aeronautics and Space Administration (NASA) scientists had instructed the astronauts on exactly how to position the mirror as part of a plan to measure the earth-to-moon distance with a laser beam.

Before lasers existed, scientists already had a fairly good idea of how far away the moon is. But "fairly good" is not good enough in science. Scientists want their measurements to be as exact as possible. Bouncing a laser beam off the mirror promised to give them more accurate figures than ever before. The experiment used the simple formula:

$$r \times t = D$$
(rate multiplied by time equals Distance)

The scientists already knew the speed of

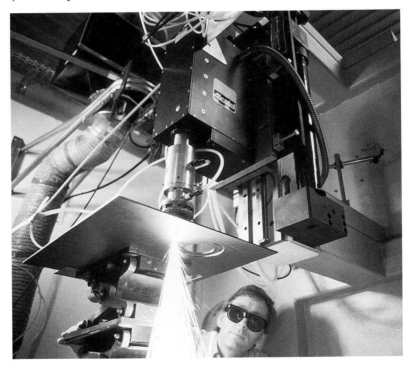

A scientist maneuvers a laser-cutting device mounted on a robot's arm. He is using the laser to cut through a metal plate.

MEASURING DISTANCES WITH LASERS

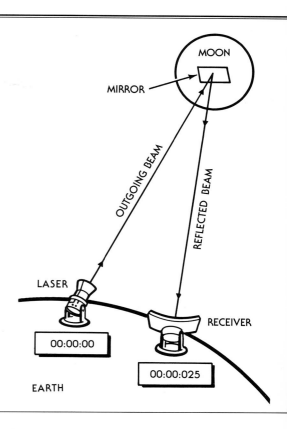

Lasers can measure enormous distances with great accuracy. A laser beam travels at a constant speed (the speed of light). The time it takes a laser beam to travel from its source, reflect off of an object, and return to the source, will indicate the exact distance between the source and the object.

light, so they knew what the rate of travel was. When the beam bounced off the mirror, it returned to earth and registered on special sensors. These recorded how long the beam took to make a round trip, and scientists then knew the time factor in the equation. After some simple multiplication, they finally had the most accurate measurement of distance possible. Knowing this has enabled them to learn much about the relationship between the earth and its natural satellite, the moon. For instance, researchers have repeated this laser-mirror experiment every year since 1969. They have found that the moon is moving away from the earth at a rate of about one-and-a-half inches (four centimeters) a year.

Measuring distance by means of lasers and mirrors works just as well on earth as it does in space. Every day, surveyors use lasers to measure the distances between houses, roads, and mountains. The laser method is more accurate and also much faster than the old surveying method, which required many calculations with poles and telescopes that had to be lined up with one another. Such measurements have also led to more exact and reliable maps. Using lasers, mapmakers have now charted almost every square mile of the earth's surface.

Toolbox Technology

Measurement is just one of several jobs that used to be associated with "toolbox" technology. Every toolbox has its yardstick, ruler, or tape measure. It also has a

PROPERTIES OF LASER LIGHT

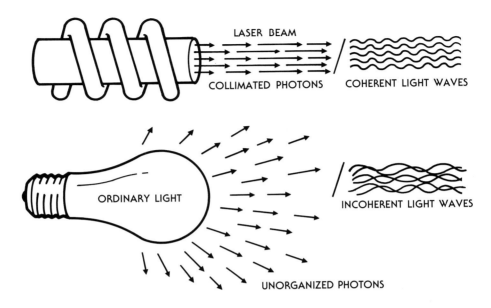

LASER BEAM

COLLIMATED PHOTONS

COHERENT LIGHT WAVES

ORDINARY LIGHT

INCOHERENT LIGHT WAVES

UNORGANIZED PHOTONS

The photons of laser light are collimated. In other words, they all travel from the laser in the same direction. Laser light is also coherent, meaning that all the waves in a laser beam have the same wave pattern. These properties make laser light more intense than ordinary light and allow it to travel long distances. Ordinary light waves scatter in all directions from their source. Also, an ordinary beam of light is incoherent, meaning that it contains waves of many different patterns, which tend to interfere with one another.

drill to bore holes and a hacksaw to cut metal. Larger toolboxes include welding equipment to join pieces of metal together. Just as lasers have come to replace the yardstick in measuring, they have also replaced the drill, the saw, and the welder. The era of the toolbox laser has arrived.

Lasers perform some toolbox jobs better and faster because of some unusual properties of laser light itself. In the first place, laser light is extremely bright, so bright that laser operators always wear protective glasses. The light is so intense because its energy is very concentrated; there are a great number of photons in a relatively small beam.

Laser light is also highly directional, or collimated. This means that all the photons travel in the same direction. They tend to stay together rather than spread apart, as the photons in ordinary light do. The farther a beam of ordinary light travels, the more it spreads out and gets dimmer. On the other hand, collimated laser light can travel a great distance without losing very much of its energy and brightness. Ordinary light could never have made it to the moon mirror and back. Laser light did.

There is one more important quality of laser light. It is coherent, or very organized. This means that the light waves are lined up with each other and moving along in step, almost like a regiment of soldiers marching in a parade. By contrast, ordinary light is incoherent. Its waves become mixed up as they move along, like the crowd of people watching the parade.

So, laser light is special. It is concentrated, directional, and organized. These three qualities combine to make the laser an extremely powerful and useful tool.

Drilling with Light

A laser is also a superdrill. This is because a laser beam can be focused into a tiny bright point. Of course, ordinary light can be focused in a similar way. A magnifying glass held up to the sun will focus the sun's rays into a tiny, very bright point. The point is also very hot, hot enough to burn a leaf or ignite a piece of paper.

Now consider collimated laser light, which is hundreds of times more directional to begin with. Such light can be focused to produce a point of light much hotter than the surface of the sun. A beam of this light can cut cleanly through

A technician adjusts the mirrors of a laser machine before conducting a metal-cutting test. Because laser light is concentrated and directional, it can cut cleanly and quickly through hard materials.

metal bars in a few millionths of a second. It can even bore a hole in a diamond, the hardest substance known. This makes the laser an excellent industrial tool.

An important use of the laser drill in industry is in the production of copper wire. The wire is formed by forcing soft copper metal into a small round hole in a diamond. The hard diamond acts like a mold, and the copper squeezes out the other end in the form of a wire. The old method of drilling the diamond holes was time-consuming and expensive. The only material hard enough to cut through a diamond is another diamond. So, workers had to use diamond drills. But diamonds are expensive. Furthermore, the drilling process took several hours, so a worker could only drill two or three holes in a workday. Today, a laser beam drills holes in diamonds at the speed of light. One worker using one laser can bore hundreds or even thousands of holes in a single hour. The same method is used for drilling holes in other gems that are used as moving parts in watches.

Burning Holes with Lasers

Oddly enough, lasers are also good at boring holes in very soft materials. Some of these materials are easily stretched or torn by ordinary drilling methods. An excellent example is the common baby bottle nipple. A laser beam burns a perfectly round hole in the top of the nipple without disturbing any of the surrounding rubber. Similarly, lasers are used to drill tiny holes in the soft plastic valves of spray cans (such as those of hair spray or glass cleaner). One such laser can punch almost seven hundred valve holes in one minute.

Lasers That Weld and Cut

Another toolbox job lasers do well is welding. The advantage of the laser over normal welding methods is similar to its advantage in other industrial areas. The laser is hotter, faster, and more accurate. It is also safer because the welder does not have to go near the hot metal.

Laser welding works on both large and small scales. On the large scale, the U.S. Navy now uses lasers to weld together huge metal parts in shipbuilding. Experts estimate that millions of dollars are saved in the welding process and millions more in reduced need for later

This high-power laser cuts a two-carat industrial diamond in half in less than a millisecond. Such lasers have become a valuable industrial tool because they are fast, accurate, and can cut through hard materials.

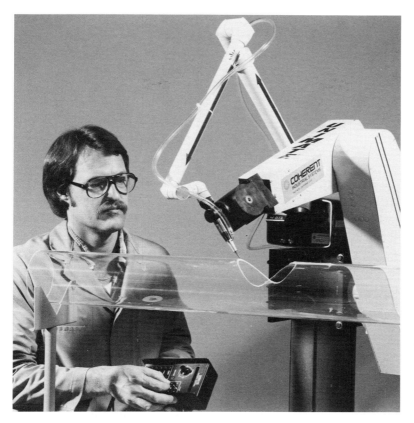

Laser devices that can cut, weld, and solder industrial materials are helping to make many production processes faster and more precise. Below, computer screens monitor the progress of a carbon dioxide laser that cuts out patterns for ceramic machine parts.

repairs. On the small scale, lasers weld the parts for tiny circuits that are used in computers, calculators, and miniature television sets. These welds, some of them microscopic, are impossible to make with normal welding tools, which only melt and destroy the delicate parts.

Today, such common items as automobile spark plugs, portable batteries, and metal braces for the teeth are routinely welded by laser beams.

Every good toolbox has a hacksaw and a pair of scissors; the saw is used to cut metal, the scissors to cut cloth. The toolbox laser can do the jobs of both. Making saw blades themselves is an excellent example of using lasers to cut metal. The old methods of producing saw blades involved many steps. During each step, people had to physically handle the

blades, and injuries were common. A laser cuts the blade out of the sheet metal in only one step. In addition, only the beam touches the metal. As long as the operator is wearing protective glasses, there is no chance for injury.

An example of the use of "laser scissors" is the cutting of patterns for clothes. A laser cloth-cutting system was designed by Hughes Aircraft, the company that employed Theodore Maiman, the inventor of the ruby laser. The system works in the following way: Pieces of cloth are laid out on a large table. The desired patterns are entered into a computer, which decides the best way to trace them out on the cloth. Next, the computer directs the laser beam to precisely cut out the pattern. Patterns for hundreds of suits can be cut in an hour. As an added advantage, the heat of the beam keeps the edges of the cloth from fraying.

The Laser as Detective

The toolbox laser can even be used to fight crime. For instance, some people are recovering stolen gems thanks to a system called laser identification. An ID marking carved into the gem is produced by a laser that creates a thin and accurate beam. This beam is so tiny that it can cleanly drill more than two hundred holes in the head of a pin. The beam is used to carve, or etch, microscopic numbers,

This carbon dioxide laser is used for welding. Faster and more accurate than traditional welding instruments, lasers are also safer. A worker can operate the laser from a distance and does not need to come in contact with the hot metal.

Garment manufacturers use laser scissors to cut patterns from cloth. Lasers cut soft materials, such as fabric and rubber baby bottle nipples, as easily as they cut hard materials.

words, or entire messages on any material, no matter how smooth or hard. This includes precious gems like emeralds and diamonds. A person's name and address can be microscopically etched on such a stone. The result is an ID marking so tiny that no one, including a thief, can detect it with the naked eye. Many other valuable items are now marked in this manner by laser beams.

Every day, several new uses are found for toolbox lasers. The devices are still rather expensive, so they are not yet normally found in home toolboxes. But this situation will surely change. As laser research continues, ways will be found to produce these tools more simply and cheaply. In the near future, a laser hanging above the basement workbench may become a common sight.

The Laser That Can Read and Write

A person who is literate is one who can read and write. A laser that can read and write can also be thought of as literate. Such a feat is possible because of the joining together of the laser and another modern supertool—the computer. Computers are able to process information thousands of times faster than human beings. For instance, hundreds of years ago, when a scientist needed to solve a complicated math problem, he or she had to do all the adding and multiplying by hand. One problem might take as long as three months to solve. Today's supercomputers can give the answer to the very same problem in only three seconds.

But scientists used to have one major problem with computers. To feed new information into the computer, people had to type the words on a keyboard, and people can only type so fast. There was a similar problem at the other end of the computing process. In order for the information that came out of the computer to be readable, the computer itself had to use a keyboard. A computer can type much faster than a person, but typing still took a considerable amount of time. So, even though a computer could process information quickly, a lot of time was wasted during the input and output stages.

A computer showing Chinese characters on its screen uses lasers to speed up the process of inputting and displaying information.

Lasers That Can Scan Type

Lasers are highly effective computer aids because they can dramatically reduce input and output times. The lasers do all the reading and writing that used to be done mechanically. Suppose that a computer programmer has a page of typed information that he or she wants to feed into the computer. Instead of typing in the information, the programmer slips the paper into a slot. Inside the machine, a tiny laser beam scans across the paper. The paper itself is white; the printed symbols are black. As it moves, the beam reflects back into a sensor that records the alternating patterns of white and black. The computer was programmed

At IBM, scientists are experimenting with new methods of using lasers to improve computer memory systems. This machine uses thousands of laser light colors to imprint information onto a disk.

earlier to recognize these patterns. The laser is able to "read" the symbols into the computer thousands of times faster than the programmer could have typed in the information.

On the output end, when the computer has information ready to be printed out, the laser once more speeds up the process. The computer "orders" the laser beam to modulate its intensity, that is, to get brighter, then dimmer, then brighter, etc., as needed. The beam has been trained to shine brighter for the symbols and dimmer for the empty spaces. The modulated beam is now scanned across a light-sensitive material, usually a TV screen or special kind of paper. The beam literally "writes" the information on the material. Again, this works many times faster than previous methods.

This and similar laser-computer sys-tems are used every day to make money transactions. For instance, most people now receive monthly bills (for electricity, gas, and credit cards) that have been printed with a laser. Many payroll checks are printed the same way. Such laser print-ers also supply people with news and information. Dozens of major newspa-pers, including *The New York Times* and the *Los Angeles Times,* use lasers to make printing plates. The laser beam etches the information on the light-sensitive plate in roughly the same way it writes on a screen or on paper.

Supermarket Lasers

Lasers that read and write have helped make many jobs easier and less boring. An excellent example can now be seen in most large supermarkets. This is the special laser scanner that reads the Universal Product Code (UPC) that now appears on most food packages in the United States. The Supermarket Institute in Washington, D.C., introduced the UPC, or bar code, in 1973. With the scanner and bar code, the checkout clerk no longer needs to press cash register buttons for every item. This not only makes the clerk's job easier but also reduces the chance for errors. A beam of light is much less likely to make a mistake than a person who is distracted, nervous, or just plain tired. In addition, the customer gets through the line and out of the store more quickly than before.

Here is how the UPC laser works: Each product in the market has a bar code (a small band of black stripes) printed on the package. A twelve-ounce box of Kellogg's Corn Flakes carries a bar code whose stripes stand for the numbers 381100. When the box is pulled across the scanner, the laser beam reads these stripes and relays the message to the computer memory bank. The computer knows (because it has been programmed to know) that these particular stripes stand for product number 381100. The computer also knows that product 381100 is Kellogg's Corn Flakes, twelve-ounce size. Smaller or larger boxes of the same product have similar but unique bar codes.

After identifying the product, the computer looks up the price, which has also been programmed in. Next, the name of the product and the price appear on a small screen that can be read by

A checkout clerk uses a laser gun to scan a sales tag for coded information. Using lasers to input information into a computer speeds up the checkout process and ensures that the customer is charged the correct amount.

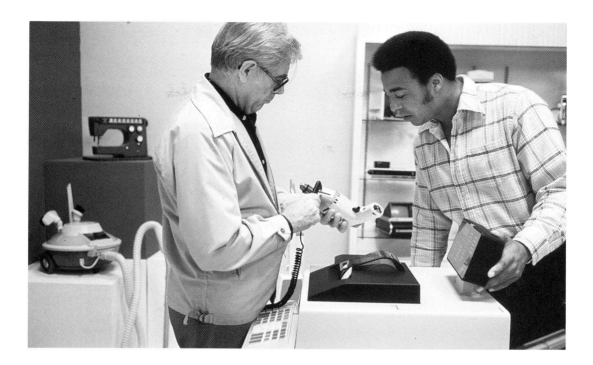

both the customer and the clerk. The total elapsed time from scanning to readout on the screen is a mere fraction of a second. The computer now moves on to the next product that the clerk scans. When all the customer's products have been scanned, the machine adds up all the prices and the total for the transaction appears on the screen.

In addition to its other advantages, the UPC system also helps the store with inventory. For instance, suppose there are one hundred boxes of Kellogg's Corn Flakes on the shelves when the store opens. Each time one of these boxes is sold, the computer records it. At the end of the day, the store manager checks the records. If the computer indicates that eighty-seven boxes of flakes sold that day, the manager knows the supply is low and reorders immediately. This obviously saves a great deal of time in walking up and down store aisles and counting cans and boxes.

When a product is scanned, a laser scanner reads the Universal Product Code printed on the object. The scanner then feeds data into a computer that describes the product being bought and tells how much it costs. The computer also reads information about size, color, and amount to help managers control their inventories.

A laser scans the bar code on a grocery item. Each type of product in a supermarket has a unique bar code.

The UPC scanner uses a helium-neon gas laser that emits a red beam. Very dependable, it is also one of the least expensive types of laser. This is important because supermarket chains buy hundreds, sometimes even thousands of the devices and could not afford the more expensive versions. The beam is powerful enough to read the bar codes but not so bright that it will hurt someone's eyes if he or she accidentally looks at it.

Machine Voices

A few more sophisticated scanners have machine voices built in so that when a product is scanned, a voice tells the customer the price. Such a system can even be programmed to say things like "Thank you. Have a nice day."

Lasers in the Office

Codes similar to the UPC are also used on magazines, books, greeting cards, and many other consumer goods. But coded symbols are not the only things laser beams can read. They can read letters and numbers too. An example is the Xerox laser facsimile system. Facsimile, or fax, machines are used to transmit photos and printed documents from one office to another. The offices can be thousands of miles apart. The only requirement is that each office has a fax machine. Weather maps are commonly sent from weather

Mirrors guide the red beam of the helium-neon laser. Supermarkets use helium-neon lasers in the scanner that reads the Universal Product Code because they are dependable and relatively inexpensive.

Left: A computerized picture scanner uses a laser to scan a picture and then print it.

Below: A laser printer works by passing a beam of light over an original document. It then burns the image into a light-sensitive material and transfers the image to paper.

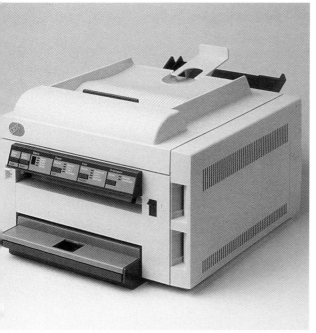

stations to TV newsrooms via fax machines.

The Xerox facsimile system works by using a helium-neon laser not unlike the ones used in supermarkets. The laser beam in the first office scans the page that will be sent (called the original document). The images contained in the reflected beam are converted into electrical energy and transmitted by wires or antennas to the machine in the second office. There, the energy is reconverted into light images, and a laser burns these images onto a light-sensitive metal drum. Finally, the drum transfers the information onto paper, and the process is complete. The whole procedure only takes about two minutes, half as long as normal facsimiles and a fraction of the time taken by hand delivery.

Optical video discs, like this one in the grasp of a robot's hand, are coded with information when laser light burns holes into the light-sensitive film that coats the discs. Later, a laser can read back the coded information in a reverse process.

Laser printers operate similarly to laser facsimile machines. The printers also use laser beams to burn images onto light-sensitive materials. These printers have revolutionized the printing and copying market in the past twenty years. There are two major reasons for this. First, they are very fast. Many print between fifty and one hundred pages per minute, while some put out as many as three hundred or more pages per minute. Second, laser printers produce unusually clear, clean copies that are excellent reproductions of the original material. These machines are extremely valuable to companies or groups that must create vast quantities of printed material. For instance, some firms send out announcements or advertisements to thousands or even millions of people in mass mailings. The laser printer, working with a computer, can even be programmed to address the envelopes.

The Beam with a Memory

The literate laser can also be used to store information. Once again, the laser works alongside or as a part of a computer system. The computer needs to have quick access to as much information as possible. The more information it can store, the larger its "memory." The laser is what "memorizes" the information. This kind of information storage is usually referred to as optical memory.

Because laser light is both collimated (directional) and coherent (organized), it can be focused down to a very small point. This point can be as little as a fraction of an inch in diameter. To create the computer memory, a small disk is coated with a smooth, light-sensitive film. Usually, this film is applied as a liquid that hardens as it dries. After converting the information to be stored into light images, the computer directs the laser to begin firing at the coated disk. The beam punches millions of tiny, microscopic holes in the film in specific patterns. The information is now stored.

Computer Playback

Later, when the computer is commanded to remember, the process is reversed. The laser scans the disk and reads back the same information. This time, however, the beam is set at lower power so it will not punch any unwanted holes in the film. After the images are read by the laser, they are transferred to the computer screen or printed out on paper so the programmer can read them.

The amount of data that can be stored by such a system is staggering. Literally billions of pieces of information can be burned into one disk, and the technology is improving all the time. With the ability to cram so much information into so small a space, the optical memory system has great potential for the future. So far, the only thing that holds the system back is that the disks do not erase, as in the case of audio and video tapes. Once the holes are burned into the film, they are permanent. Each disk can be used only once. This means that it is impossible to write over existing data to correct or update information. To do so, a com-

This man is working with an optical computer to produce graphic designs. Optical computers use photons, rather than electrons, to store and retrieve information. These computers have the potential to be thousands of times faster than conventional computers. COURTESY OF AT&T ARCHIVES

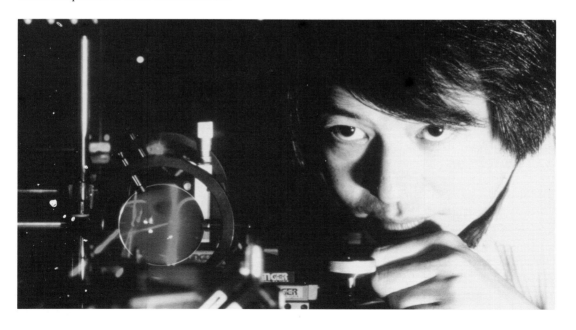

Researchers David Miller (left) and Jill Henry of AT&T Bell Laboratories adjust equipment used to build the first photonic switching chip, a component of the optical computer. COURTESY OF AT&T ARCHIVES

pletely new disk has to be programmed, and that takes time. Also, the disks are expensive to produce. Researchers are working to build an erasable disk, and it seems like only a matter of time before they will succeed.

Computing at Light Speed

For several years, the laser has joined in a useful working partnership with the computer. But the laser still only reads, writes, and memorizes for the computer. Some scientists think the laser could go further and bring about a drastic change in the way the computer is designed.

The computer itself consists of wires, chips, connections, and other parts, through which electrical signals flow. Experts point out that, in the larger supercomputers, sometimes too many pieces of information are trying to get to the same place at the same time. Due to the limitations of the machine parts themselves, the information bits can only move so fast. As a result, bottlenecks form. These are like miniature traffic jams, only with bits of data instead of cars.

Another problem with computers has to do with electrical "noise." The electrical signals must move along wires, many of which are close together. When the signals interfere with one another, they create noise. Like the bottlenecks, noise interference makes the computer work less efficiently.

The laser might be able to eliminate both of these problems. Alan Huang, a researcher at the AT&T Bell Labs in Holmdel, New Jersey, wants to build an optical computer. This device would use laser light instead of electricity to process the information. In the first place, a laser beam could carry millions of signals without once touching a physical connection. Thus, bottlenecks would be eliminated, and much more information could flow through the computer. In addition, unlike wires, laser beams can pass near or even right through each other without any interference. That would quiet the noise problem.

Huang admits that there are still many technical problems to be worked out before an optical computer can be built. But most people agree that the possible benefits are definitely worth whatever time and effort are required. The optical computer, says Huang, will "leave today's fastest supercomputers in the dust."

The Light That Talks

By 1960, the beginning of the laser era, science had given humanity several reliable communications devices. These included the telegraph, the telephone, radio, and television. At the time, the communications satellite was just getting off the ground. Each of these inventions revolutionized the world in its own particular way. The laser, however, has made possible a communications revolution far larger than all the others combined. The main reason for this giant leap forward in information exchange is a physical quality of light itself. It can carry a great deal of information.

The amount of information light can carry depends on its frequency. Imagine going to the beach two days in a row. On the first day, the ocean waves are long and lazy; their crests are about fifty feet apart. On the second day, the situation is much different. The waves are now much shorter and more energetic. Their crests are only five feet apart. Obviously, there are more waves (ten times more to be exact) breaking per minute on the second day than on the first. Because the waves on day two are more frequent, they are said to have a higher frequency.

Waves of the different types of radiation (radio, microwaves, or light, for example) behave somewhat like the ocean waves. The lower frequency radiation waves are long and lazy. The higher frequency waves are short and energetic. The important point here for communications is that the higher the frequency, the more information that can be carried.

Consider that the telephone transmits the human voice at a frequency of about three thousand waves, or cycles, per second. That sounds like a large number of waves until it is compared to a television signal. Television transmits at a frequency of about fifty million cycles per second. Obviously, a lot more information can be carried by a television signal than by a telephone signal. In fact, that is why the telephone can only transmit a voice, whereas television broadcasts voice and picture.

But even the frequency of television signals is small compared to beams of light. Visible light frequencies range between four hundred trillion and eight hundred trillion cycles per second. That means that light has the capacity to carry more than one million times as much information as television. In communications, the amount of information exchanged is the most important factor. It is no wonder then that the laser which uses light to transmit information, has been so revolutionary.

Laser communications work in two basic ways. One way involves transmission directly through the atmosphere (or through space). The second way works in combination with the science of fiber optics.

Signals Through the Air

As discussed earlier, the direct transmission of a laser beam to a mirror on the

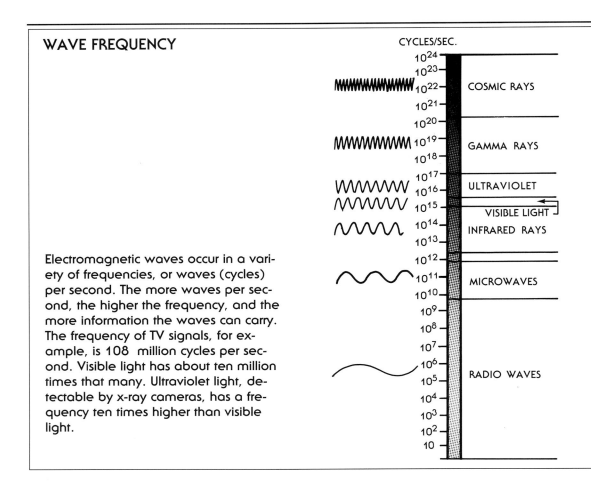

WAVE FREQUENCY

CYCLES/SEC.

- 10^{24}
- 10^{23}
- 10^{22} — COSMIC RAYS
- 10^{21}
- 10^{20}
- 10^{19} — GAMMA RAYS
- 10^{18}
- 10^{17}
- 10^{16} — ULTRAVIOLET
- 10^{15} — VISIBLE LIGHT
- 10^{14} — INFRARED RAYS
- 10^{13}
- 10^{12}
- 10^{11} — MICROWAVES
- 10^{10}
- 10^{9}
- 10^{8}
- 10^{7}
- 10^{6}
- 10^{5} — RADIO WAVES
- 10^{4}
- 10^{3}
- 10^{2}
- 10

Electromagnetic waves occur in a variety of frequencies, or waves (cycles) per second. The more waves per second, the higher the frequency, and the more information the waves can carry. The frequency of TV signals, for example, is 108 million cycles per second. Visible light has about ten million times that many. Ultraviolet light, detectable by x-ray cameras, has a frequency ten times higher than visible light.

moon enabled scientists to measure the distance between the earth and moon. The same basic principle could be used to communicate with astronauts on a moon base. In this case, the beam would carry stores of information and would not bounce back to earth. A receiver in the base would pick up the beam, and a computer would decode it.

Getting back down to earth, the very same procedure is already in common use in many cities. Several businesses have set up systems to flash laser beams from building to building. Some park services also use this idea. Sometimes a ranger is stationed at a base on a remote mountain. Installing telephone lines such a great distance through rugged territory could be too expensive. The ranger has a radio, but if a large amount of information must be sent, the laser is a better choice. A fellow ranger at the main headquarters sends a communications beam to a receiver at the mountain base. A small computer in the base decodes the beam and stores the information. Later, when he or she has the time, the ranger can display and study sections of the message on the computer screen.

Unfortunately, light does not travel well through the atmosphere. The individual molecules of air tend to absorb some of the photons as they travel along. The farther the light goes, the dimmer it gets. Alexander Graham Bell had discovered this problem when he was trying to build his photophone in 1880. Scientists in the early 1960s also noticed the problem when they began experimenting with communications lasers. Even bright, coherent laser light is weakened by traveling through the air. As a result, direct transmission of laser light has been limited mainly to short-range situations, like communication with the mountain base. (Lasers carrying information to the moon can be considered an exception because most of the trip is through space, where there are no air molecules like those found in the earth's atmosphere.)

Yet the fact that laser light could carry so much information excited researchers. They realized that, among other things, it could carry huge numbers of telephone or television signals in a single beam. All scientists had to do was figure out a way to bypass the atmosphere. Soon, they realized that the answer had been staring them in the face for many years.

The Coming of Fiber Optics

In 1934, an inventor named Norman R. French patented an idea for a device called the light pipe. French proposed taking a hollow pipe and lining it with a reflective material. If someone shined a light into the pipe, the rays might bounce off the inner surface and keep going through the tube. French's pipe was not for sending information. He was trying to find a way to carry illumination from one room to another. But scientists in the 1960s believed that the light pipe idea could be adapted to the field of laser communications.

The researchers quickly realized that the pipe they needed could not actually be hollow because then it would still contain air. The air would absorb the light as it did in the atmosphere. Also, the signal would lose some of its power because small amounts would be absorbed by the lining itself. And there was another problem. Light travels in straight lines, and the pipe would have to curve now and then to avoid obstacles (especially since it would be placed underground). There had to be a way to keep the curves from blocking the beam.

In 1966, two British researchers, Charles Kao and George Hockham, suggested that thin glass fibers might be able to transmit light over short distances. Other scientists quickly picked up the

A sphere of green light glows from the end of a fiber-optic cable. In the background, the cable is wound around a spool. COURTESY OF AT&T ARCHIVES

LASERS AND FIBER-OPTIC COMMUNICATION

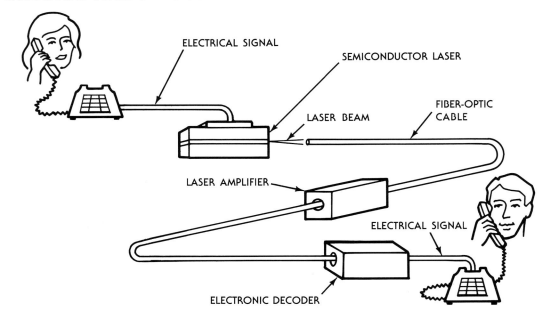

ELECTRICAL SIGNAL

SEMICONDUCTOR LASER

LASER BEAM

FIBER-OPTIC CABLE

LASER AMPLIFIER

ELECTRICAL SIGNAL

ELECTRONIC DECODER

Optical fibers are thin strands of glass through which a laser beam can travel for several miles. Since a laser sends signals on light waves, a single optical fiber can carry as much information as hundreds of heavy copper wires, which carry electrical signals. When electrical signals travel through copper wires, they are quickly weakened. Devices called repeaters are needed about every mile to strengthen the electrical signal. In a fiber-optic system, laser amplifiers are needed only every six or seven miles to strengthen the light signal.

idea, and in 1970 Robert Maurer of the Corning Glass Works in Corning, N.Y., constructed the first long-distance optical fiber. The science of fiber optics was born.

The fiber idea eliminated some, but not all, of the problems of the light pipe. The system uses glass fibers only a fraction of an inch in diameter. The fibers, which make up the core, are stuffed inside a small cable that is lined with a material known as the cladding. This is an extremely reflective type of glass that makes most of the stray photons bounce back into the core. The cladding eliminates the problem of the beam not being able to move around curves. As long as the curves are not too sharp, the beam hits the cladding at an angle, then continues on. There is no air in the cable to absorb the photons, but a few do get through the cladding and eventually weaken the beam. Scientists realized early on that they would need to reamplify the light beams every now and then. So, today's optical cables have small stations set up at intervals to boost the signals.

Conversing by Light Beams

To transmit the light, a laser beam carrying information is flashed into a fiber. The light travels along (at light speed, of course) until it encounters a boosting device. There, it gets amplified and continues on. This process might be repeated a few or many times, depending on the distance involved. At the end of the cable, a receiver catches the light beam, and the signal is decoded.

In Maurer's 1970 experiments, the system worked well for a distance of only about 1/3 of a mile (about 500 meters). After that, too many photons got absorbed, and the signal faded. Today, some optical fibers carry light more than 26 miles (40 kilometers) before the signal has to be reamplified. Lab experiments have shown that this distance may be increased to almost 133 miles (200 kilometers) in the near future.

Of course, one of the advantages of laser fiber optics is that several fibers can be wrapped inside one cable. This means that each cable contains many laser beams, each carrying billions of bits of information. This and other advantages mentioned make the optical system clearly superior to earlier systems.

For instance, the conventional telephone cable used wires to transmit conversations. Obviously, to carry many conversations, there had to be many wires in a single cable. Thus, the cable was thick, heavy, and difficult to install. Also, since the metal wires had to be packed so closely together, the separate signals interfered with each other and produced electrical noise.

By contrast, the optical telephone

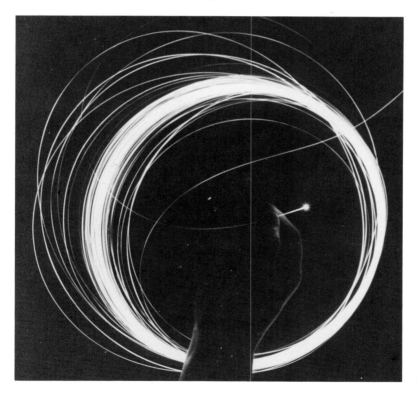

These loops of glass fiber are as thin as hair and illuminated by laser light. Many fibers can fit inside one cable and can carry billions of bits of information. COURTESY OF AT&T ARCHIVES

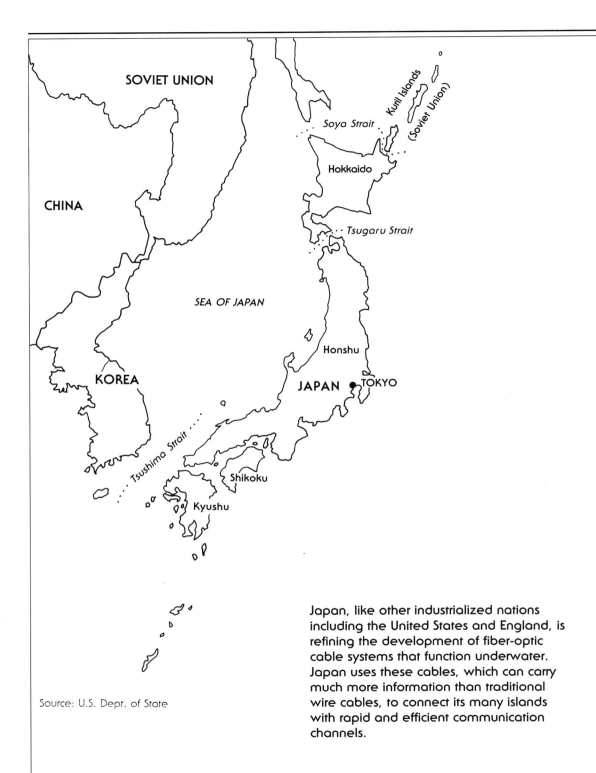

SOVIET UNION

Kuril Islands
(Soviet Union)

Soya Strait

Hokkaido

CHINA

Tsugaru Strait

SEA OF JAPAN

Honshu

KOREA

JAPAN ●TOKYO

Tsushima Strait

Shikoku

Kyushu

Japan, like other industrialized nations including the United States and England, is refining the development of fiber-optic cable systems that function underwater. Japan uses these cables, which can carry much more information than traditional wire cables, to connect its many islands with rapid and efficient communication channels.

Source: U.S. Dept. of State

system uses a much thinner, lighter cable that is easier to install. Beams of light do not interfere with each other, so there is no noise in the system. A large conventional telephone cable could carry as many as a few thousand conversations at one time. Fiber-optic cables now exist that can carry several hundreds of thousands of conversations at one time. Moreover, scientists predict this number will exceed one million in the very near future.

Several countries, including the United States, Japan, and England, are now using fiber-optic cables that travel underwater. For instance, England and France, which are separated by the English Channel, exchange huge amounts of information daily. Conventional cables using wires already stretch across the channel. But these will soon be totally outdated by the optical cables, which can carry dozens of times more information.

Similarly, Japan, a country made up of many islands, is connecting all of these islands via undersea optical cables. Also, a new transatlantic cable that will connect North America and Europe is already being installed by American engineers.

The Fiber-Optic Society

Since a laser beam can carry high-quality television signals, more and more companies are installing cable TV lines that use laser fiber optics. Experts estimate that, by the year 2000, most American homes will be linked by a complex network of optical cables. This will bring better and more reliable reception and, of course, more channels. It will also allow any one house to be hooked up directly to someone who provides a specific service. For example, a person could run a daily exercise clinic from his or her living room. Ten, twenty, or more subscribers (those paying for the service) could tune in and participate within the comfort of their own homes.

Also, the fiber-optic system will connect individual homes to libraries and other storehouses of information. A person will be able to ask a librarian to enter a book, magazine article, or even a movie into the library computer. A fiber-optic cable will then carry the desired information to a screen in the person's home. Systems like the ones described here are already in use in several experimental areas (including an entire town in Japan that has been wired for fiber optics). As demand increases, the systems will be installed in more cities and towns.

It should be noted that fiber optics can work without lasers. When the dis-

Fiber optics can transmit a lot of information at high quality. They are often used in cables to carry TV signals that provide better reception.
COURTESY OF AT&T ARCHIVES

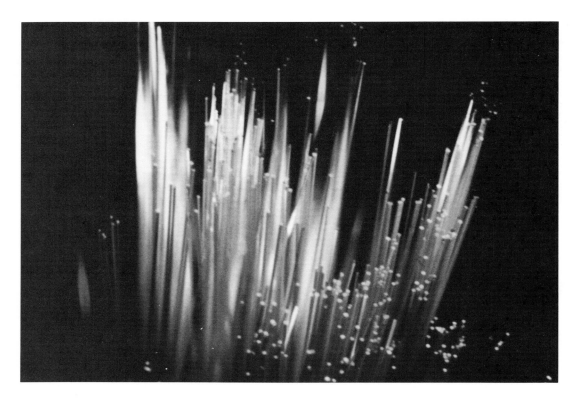

Bursts of light glow at the ends of these fiber-optic cables. Laser communications using fiber optics will become even more important in the future as people begin to demand faster access to more information. COURTESY OF AT&T ARCHIVES

tance the fibers have to travel is very short and/or very little information needs to be transmitted, more conventional types of light will suffice. An example is the fibered automobile. Most major car manufacturers now use at least some optical fibers to replace wires in cars. Eventually, the entire electrical systems of cars will work through fibers. Fibered cars will use light sources like the LED (light-emitting diode), which uses a crystal to produce a very intense version of ordinary light.

However, the bulk of fiber-optic systems uses lasers, and this trend will continue into the foreseeable future. The choice of the laser for future communications is inevitable because laser light can carry vast amounts of information. It has been estimated that more than 100,000,000 television channels might be transmitted using the frequencies in the spectrum of visible light. Even if only one-tenth of 1 percent of this total is ever used, that is still 100,000 channels. Only laser light will be able to carry that much information.

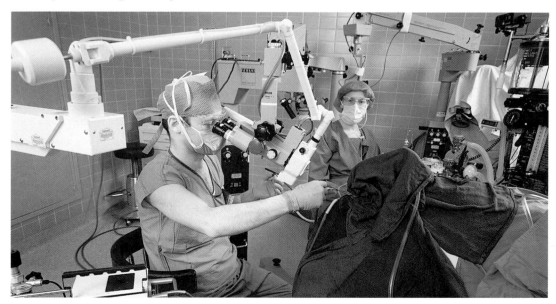

■■■■■■■■ CHAPTER **7**

The Laser and Medicine

One of the biggest surprises in the early days of lasers was that these tools of light could be used in the science of medicine. No one had thought that lasers might be able to heal. But doctors and medical researchers quickly began to suggest possible uses.

Some doctors pointed out that there are surgical operations that are difficult to perform with the conventional scalpel. Perhaps the laser could be used instead. Experiments showed that a finely focused beam from a carbon dioxide gas laser could cut through human tissue easily and neatly. The surgeon could direct the beam from any angle by using a mirror mounted on a movable metal arm. Several advantages quickly became apparent. First, the light beam is consistent, which means that it gives off the same amount of energy from one second to the next. So, as long as the beam is moving along, the cut it makes, called the incision, does not vary in depth. Using a scalpel, a doctor can accidentally make part of the incision too deep.

Bloodless Surgery

A second advantage of the surgical laser is that the hot beam cauterizes, or seals

Surgeons often use lasers instead of scalpels to perform delicate operations. The thin beam of laser light can make a precise incision in human tissue with more accuracy than a surgeon's knife.

off, the open blood vessels as it moves along. However, this works well mainly for small vessels, such as those in the skin. The doctor still has to seal off the larger blood vessels using conventional methods.

Because cells in human tissue do not conduct heat very well, the skin or any other tissue near the laser incision does not get very hot and is not affected by the beam. This advantage of laser surgery is very helpful when a doctor must operate on a tiny area that is surrounded by healthy tissue or organs.

It should be pointed out that the "laser scalpel" is not necessarily the best tool to use in every operation. Some doctors feel that while the laser is useful in some situations, it will never totally replace the scalpel. Others are more optimistic and see a day when more advanced lasers will make the scalpel a thing of the past. For the present, lasers have proved to be most effective in operating on areas that are easy to reach—areas on the body's exterior. These include the skin, mouth, nose, ears, and eyes.

This is not to say that using a laser for internal surgery is impossible. In fact, doctors have shown remarkable progress in this area. Of course, in order to be able to direct the laser beam, the doctor must be able to see inside the body. In some cases, this is a simple matter of making an incision and opening up the area to be operated on. But there are situations in which this step might be avoided.

Cleaning Arteries with Light

For instance, lasers seem to be one way of cleaning plaque from people's arteries. Plaque is a tough, fatty substance that

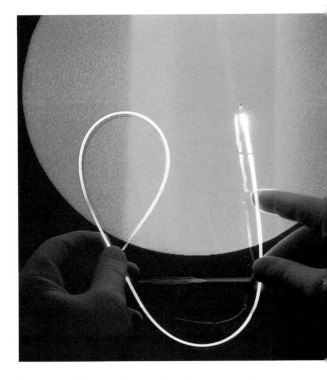

A doctor points the hot tip of a laser at a section of an angiogram, a kind of X-ray photograph of blood vessels, to show where surgery will occur.

builds up on the inside walls of some people's arteries. Eventually, the vessels can get so clogged that blood does not flow normally. The result can be a heart attack or stroke, both of which are serious and sometimes fatal. The normal method for removing the plaque involves opening the chest and making several incisions. This is a long and sometimes risky operation. It is also expensive and requires weeks for recovery.

As an alternative, a laser beam could burn away the plaque. The key to making this work is the doctor's ability to see inside the artery and direct the beam. Here is another area in which fiber optics and lasers are combined into a modern wonder tool. An optic fiber that has been connected to a television camera can be

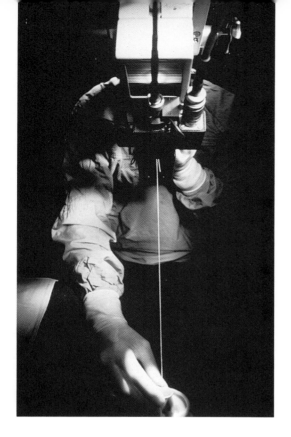

Using an argon laser carried by a fiber optic, a doctor performs surgery on a patient's inner ear.

is needed (except for the small one in the artery to insert the fibers). There is also little or no bleeding, and the patient can enjoy total recovery in a day or two.

But this method does have some disadvantages too. When the laser beam fires at the plaque, it must be aimed very carefully. A slight miss could cut through the wall of the artery and cause serious bleeding. The patient's chest would then have to be opened up after all. Another problem involves small pieces of burnt debris from the destroyed plaque. If these enter the bloodstream, they can cause blockages in smaller blood vessels, bringing further complications.

Until these problems are overcome, laser angioplasty should be thought of as experimental. However, research continues and the treatment, no doubt, will become common.

Restoring the Miracle of Sight

inserted into an artery. The fiber now becomes a miniature sensor and the doctor can see inside the artery. A second fiber is inserted to carry the bursts of light that will burn away the plaque.

The technique works in the following way: The two fibers are inserted into a vein in an arm or leg and moved slowly into the area of the heart and blocked arteries. When the fibers are in place, the laser is fired and the plaque destroyed. The exhaust vapors are sucked back through a tiny hollow tube that is inserted along with the optical fibers. When the artery has been cleaned out, the doctor removes the fibers and tube, and the operation is finished. This medical process is known as laser angioplasty. It has several obvious advantages. First, no incision

Some of the most remarkable breakthroughs for medical lasers have been in the area of ophthalmology. This is the study of the structure and diseases of the eye. One reason that laser beams are so useful in treating the eye is that the cornea is transparent. The cornea is the coating that covers the eyeball and admits light into the interior of the eye. Since it is designed to admit ordinary light, the cornea lets in laser light just as well and remains unaffected by the beam.

The laser is very useful in removing blood vessels that can form on the retina. The retina is the thin, light-sensitive membrane at the back of the eyeball. It is on the retina that the images of the things the eye sees are formed. Damage to the retina can sometimes cause blindness.

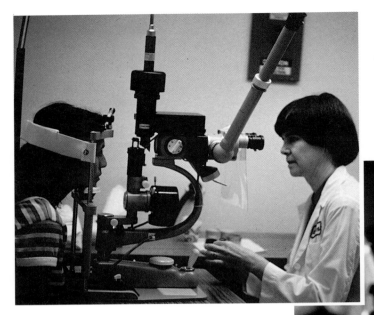

The doctor and diabetic patient prepare for eye treatment using an argon laser. Diabetes can often cause a type of blindness in which tiny blood vessels form on the eye's retina. A laser can effectively burn away these vessels and restore sight.

The most common form of blindness in the United States is due to a disease known as diabetes, which is characterized by high levels of blood sugar. In some advanced cases of diabetes, hundreds of tiny extra blood vessels form on the retina. These block light from the surface of the membrane. The result is partial or total blindness.

The laser most often used in the treatment of this condition has a medium of argon gas. The doctor aims the beam through the cornea and burns away the tangle of blood vessels covering the retina. The procedure takes only a few minutes and can be done in the doctor's office. In a nationwide study, doctors reported that the laser method worked in about half the cases treated.

The laser can also repair a detached retina, or one that has broken loose from the rear part of the eyeball. Before the advent of lasers, detached retinas had to be repaired by hand. Because the retina is so delicate, this was a very difficult operation to perform. Using the argon laser, the doctor can actually "weld" the torn retina back in place. It is perhaps a strange coincidence that Gordon Gould, one of the original inventors of the laser, later had one of his own retinas repaired this way.

Another condition that affects the eye is glaucoma, a condition characterized by a buildup of fluid in the eye. Normally, the eye's natural fluids drain away a little at a time, and the eye stays healthy. In eyes impaired by glaucoma, the fluid does not drain properly, and the buildup affects vision. Blindness can sometimes result.

In some cases, drugs can be used to treat glaucoma. If the drugs fail, many doctors now turn to the laser to avoid conventional surgery. The laser punches

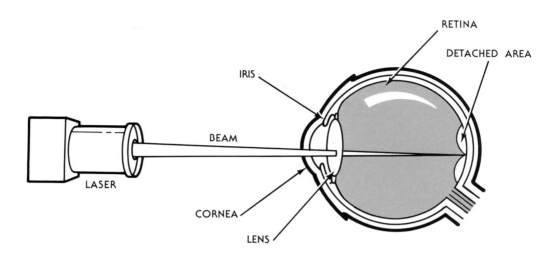

USING LASERS FOR EYE SURGERY

RETINA

DETACHED AREA

IRIS

BEAM

LASER

CORNEA

LENS

The laser works like a sewing machine to repair a detached retina, the membrane that lines the interior of the eye. The laser beam is adjusted so that it can pass harmlessly through the lens and focus on tiny spots around the damaged area of the retina. When it is focused, the beam has the intensity to "weld" or seal the detached area of the retina back against the wall of the eyeball.

a hole in a preplanned spot, and the fluid drains out through the hole. Again, the treatment can be performed in a doctor's office instead of a hospital.

Zapping Birthmarks

Recently, lasers have come to be used in removing port-wine stains. These are reddish purple birthmarks that appear on about three out of every one thousand children. The stains can mark any part of the body but are most commonly found on the face and neck.

The medical laser is able to remove a port-wine stain for the same reason a military laser is able to flash a message to a submerged submarine. Both lasers take advantage of the monochromatic quality of laser light, its ability to shine in one specific color. The stain is made up of thousands of tiny, malformed blood vessels that have a definite reddish purple color. This color very strongly absorbs a certain shade of green light. In fact, that is why the stain looks red. It absorbs the green and other colors in white light but reflects the red back to people's eyes.

To treat the stain, the doctor runs a wide, low-power beam of green light across the discolored area. The mass of blood vessels in the stain absorbs the energetic laser light and becomes so hot that it is actually burned away. The surrounding skin is a different color than the stain.

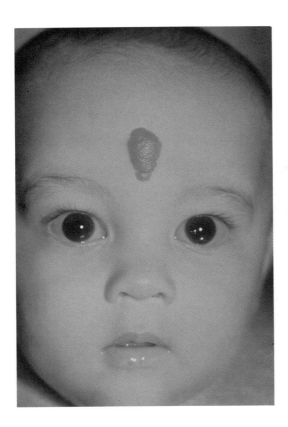

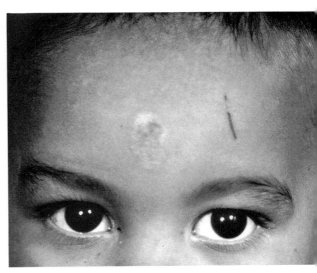

Left: This child was born with a port-wine stain, a birthmark that usually appears on the face or neck.

Below: The discoloration disappeared after a green laser light burned the birthmark away.

The skin absorbs only small amounts of the beam and remains unburned. Of course, the burned areas must now heal. During this healing process, minor scarring sometimes occurs, and researchers are now trying to find ways to prevent this.

A similar method has sometimes been successful in removing tattoos. A tattoo is formed when very strong dyes are injected with needles into a person's skin. Someone who has been tattooed may decide later in life that he or she does not want the tattoo any more. In the past, the only way to remove these designs involved surgery or burning off the tattoo with acid. Luckily, the laser offers an alternative to such extreme measures. The beam bleaches the dyes in the tattoo without burning the surrounding skin. As with port-wine stains, some light scarring is possible.

The Age of Painless Dentistry

Believe it or not, lasers are beginning to make some people actually look forward to going to the dentist. Laser dentistry is a rapidly expanding branch of medicine. No one enjoys having a cavity drilled. It usually requires an anesthetic (a pain-killer like novocaine) that causes uncomfortable numbness in the mouth. Also, the sound of the drill can be irritating, even sickening to some people.

Some dentists are now using a Nd-YAG laser (one that uses a crystal for its lasing medium) to replace the drill. The laser treatment takes advantage of the simple fact that the material that forms in a cavity is much softer than the enamel, which is the hard part of a tooth. The laser is set at

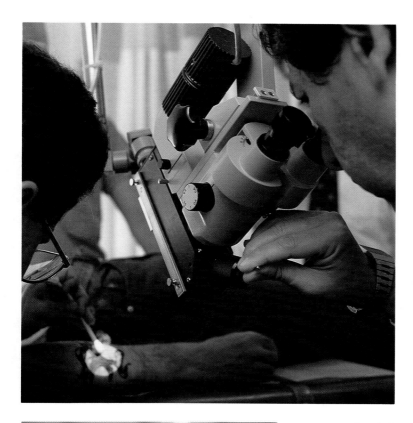

Left: Doctors use a laser to remove a tattoo. The light bleaches the dyes in the tattoo until the colors disappear.

Below: A red helium-neon laser beam is aimed into the patient's mouth. Dentists are beginning to use lasers instead of drills to treat cavities.

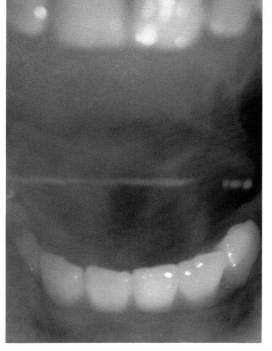

a power that is just strong enough to eliminate the decayed tissue. But the beam is not strong enough to harm the enamel. When treating a very deep cavity, bleeding sometimes occurs. The laser beam seals off blood vessels and stops the bleeding.

The most often asked question about treating cavities with lasers is: Does it hurt? The answer is no. Each burst of laser light used lasts only thirty-trillionths of a second. That is much faster than the amount of time a nerve takes to trigger and start hurting. In other words, the beam would have to last 100,000,000 times longer in order to cause any pain. So, this sort of treatment requires no anesthetic.

There are literally hundreds of medical uses for the laser. Only a few have been discussed here. Of course, there are many medical conditions a laser cannot help.

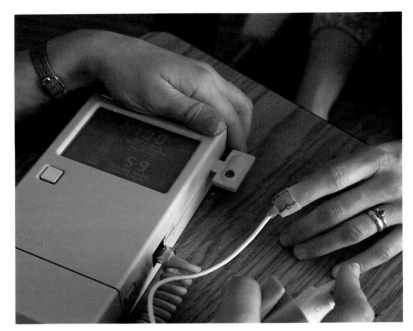

Left: A portable pulse oximeter takes a patient's pulse using a laser. The patient wears a laser wrapped around her finger underneath a bandage.

Below left: A laser beam destroys a kidney stone.

Below right: This is a computer-generated picture of a chromosome, a microscopic piece of material in a cell that contains genetic material. The arrow points to where a laser has deleted a piece of the chromosome.

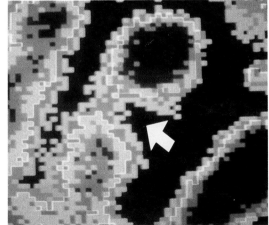

Even when it can help, a doctor may have a good reason for choosing a different method. While the laser is a marvelous medical tool, it cannot cure every ill.

Yet the world has seen probably only a small fraction of the laser's potential.

Consider that this supertool has only existed for about thirty years. Try to imagine what the healing laser will do in another thirty years.

The Laser and Entertainment

So far, this book has described many of the useful things scientists have found for the laser to do. The device can measure, cut, drill, weld, read, write, send messages, solve crimes, carry telephone conversations, burn plaque out of arteries, and perform delicate eye operations. The laser is an extremely practical tool.

But does everything in life have to be practical? Human beings are intelligent creatures and have used their intelligence to build useful tools and a technical civilization. But another side of intelligence is the ability to appreciate beauty. Humans are perhaps the only creatures in the world who want to look at something or do something simply for the fun of it. As a result, men and women have always sought ways to entertain themselves and each other.

Light itself can be beautiful. The sight of a deep red sunset or multicolored rainbow can inspire feelings of happiness, romance, and even awe. For centuries, artists have tried to reproduce light's beauty in paintings. In addition, inventors have given artists mechanical tools such as the camera, which uses light to create art that is entertaining as well as beautiful. Since the laser produces a special kind of light, people realized its potential to create special kinds of art and entertainment.

Painting with Light

In the late 1960s, artists began to use lasers to produce "light paintings." These took

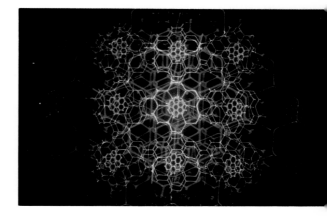

Colored lasers can be manipulated to produce intricate and beautiful patterns. Lasers are not only useful tools but they are also becoming an important artistic medium.

the form of one-time performances in which an artist flashed laser beams in various ways to create visually striking patterns. The beams might be bounced off mirrors placed in preplanned positions. Sometimes the mirrors would be attached to the artist, who would move about, reflecting the rays against walls, glass objects, or into tanks filled with liquid. Another variation involved bouncing the beams against clouds of machine-made fog. Usually, the performance was done to music. The effects of these displays could often be exciting to watch, especially at night when the beams glowed brightly against the dark sky.

Unfortunately, not many artists can afford the equipment necessary for light paintings. So, such shows are now more often staged by organizations that put on public performances for fees. Sometimes, large fairs and celebrations present displays of laser art. An early example appeared in the Pepsi-Cola Pavilion at Expo '70 in Osaka, Japan.

Lasers with a Beat

The laser artists at Expo '70 set up rotating mirrors and wired them to equipment that played music. They aimed four colored laser beams—red, yellow, green, and blue—at the mirrors. When the music played, the sounds traveled through the wires and caused the mirrors to spin at different speeds. The mirrors bounced the beams around the room in complex patterns, sometimes to the beat of the music. Most of the spectators commented that the show was strikingly beautiful.

A much more spectacular display of laser art occurred during the United States bicentennial celebration in 1976. The performance was staged at the Washing-

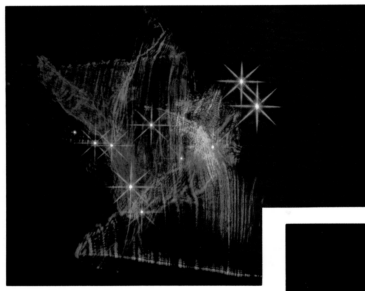

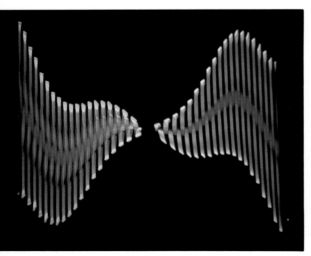

These three examples of laser art demonstrate the versatility and brilliance of laser light.

ton Monument. An audience of four million people watched the show, and the beams could be seen twenty miles (thirty kilometers) away. Other laser artists staged two such large-scale presentations in 1980. One helped celebrate the city of Boston's 350th birthday. The other was set at the party for President Ronald Reagan's inauguration.

In these major laser shows, the light beams could be considered the main attraction; the music supported the visual display. It soon became clear to people in the music business that the reverse could work just as well. Thus, it has become common to witness laser shows at music concerts, especially rock concerts. In such cases, the music performance is the main attraction, while the light show takes on the supporting role. Many well-known recording groups have staged these light shows at their concerts, including The Who, Def Leppard, and Foreigner.

From time to time, there have been

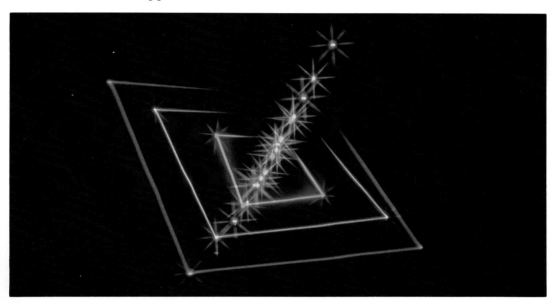

Laser light concerts are often shown in planetariums. These dome-shaped theaters allow the images to surround the audience.

some questions about safety during rock concert laser shows. Some of the early displays allowed the beams to shine into the audience. This is potentially dangerous. The beams are not powerful enough to burn a person's skin; however, if a beam shines directly into someone's eye, a permanent blind spot can form. Because of this danger, some countries have established strict rules about how lasers can be used in concerts.

The Laser Disc Revolution

While the laser continues to thrill people in large visual shows, it also entertains them on a small scale in their homes. In the late 1970s and early 1980s, a revolution in viewing and listening technology began. First came the videodisc player, which plays movies and other shows on a television screen. The disc is encoded with the visual information (the movie) in roughly the same way that computer storage discs are encoded. A laser beam burns patterns into a film that covers the disc. Later, a small laser inside the player scans the disc and relays the picture to the screen. The picture produced by a videodisc is brighter and sharper than the one produced by a videotape.

Unfortunately, the first videodisc players that hit the market had many problems. A great many had not been built well, and people returned them. Furthermore, the companies that built them did not make and supply a wide enough variety of discs to satisfy customers. Many more titles existed on tape, and tape

Videodiscs ushered in a revolutionary technology. Although videodiscs were unsuccessful in the marketplace, they provided the science that created audio discs, or compact discs, which have been a huge success. These discs have begun to make conventional records and tapes obsolete.

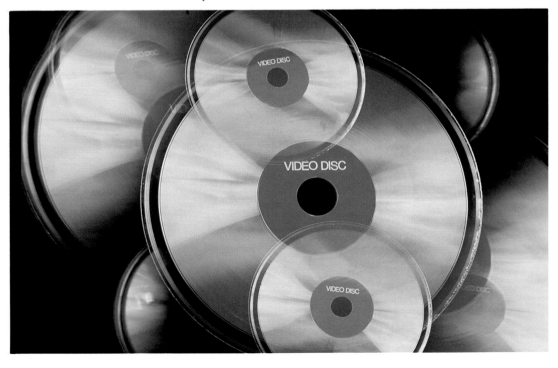

THE CD PLAYER

One side of a compact disc has a reflective coating in which a pattern of pits has been etched. As shown in the enlargement below, a laser beam reflects off these pits onto a light-sensitive transmitter. The transmitter converts the pattern of reflections to electronic signals, which are converted to sound.

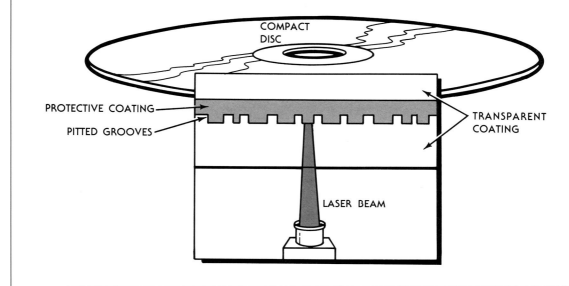

COMPACT DISC

PROTECTIVE COATING

PITTED GROOVES

TRANSPARENT COATING

LASER BEAM

players themselves seemed more reliable, so videodiscs did not catch on with consumers.

Soon after videodiscs appeared, laser audio discs (usually called compact discs) hit the market. These did catch on with the public and have already begun to replace traditional long-playing records. One reason for the success of the compact disc is its excellent sound reproduction. In a phonograph, a needle comes in direct contact with the carved grooves on the surface of the record. The more the record is played, the more the grooves wear down. In addition, they can hold only a certain amount of musical information.

In a compact disc player, on the other hand, only a beam of light touches the surface of the disc. So, barring accidents or misuse, the discs do not wear out. Also, the disc can carry more information than a record, and the sound is sharper and more realistic.

The other reasons for the success of compact discs are the same as for the success of videotapes over videodiscs. The compact disc players proved to be largely reliable, and few customers returned them. In addition, manufacturers quickly made tens of thousands of titles available. In fact, many titles appeared on discs that could not be found on records.

Recently, researchers have managed to greatly improve videodisc technology. The new players are far more reliable than the old ones. Also, the disc-produced pictures are more sharp and realistic than ever. In an effort to attract new customers, manufacturers have offered deluxe editions of movies. These often include restored footage, scenes that appeared in the original film but got lost over the years. A notable example is the recent release of Orson Welles's classic *The Magnificent Ambersons*. When Welles made the film in the early 1940s, studio bosses did not like it and ordered editors to change, shorten, or take out certain scenes. The film was then released to the public. The disc version has replaced the missing scenes where possible. Makers of videodisc players are hopeful that this second marketing plan will be more successful than the first.

Three-Dimensional Images

Lasers can also create a form of light-produced art called holography. This field deals with a type of photography that creates three-dimensional pictures. By contrast, a standard camera produces pictures that are only two-dimensional.

Holography began to develop in the late 1940s, quite a while before lasers appeared. The basic idea was to shine two separate beams of light at a sensitive sheet of photographic film. One beam would bounce off the object being photographed. The other would travel a different path, and both beams would reach the sheet of film at the same time. Once exposed by the light, the film itself became the hologram. Later, when a person shined a third beam at the hologram, a three-dimensional picture of the object was supposed to be visible.

The idea made sense as a theory, but it was very difficult to construct a working model. One problem was that the light in the beams had to be coherent, moving along with all the waves in step. Another problem was that both beams had to be monochromatic, of one color. Producing two identical beams with these properties was an almost impossible task at that time. Researchers tried all kinds of light sources, but none worked very well and progress in holography was slow all through the 1950s.

Then, in 1960, Theodore Maiman built his ruby laser and holography received a sudden boost. Researchers now had a light source that was bright, coherent, and monochromatic. They found that they could produce two identical laser beams by passing a single laser beam through a device called a beam splitter. These beams bounced off a series of mirrors to reach the photographic film.

Today, almost everyone has seen a hologram at one time or another. The three-dimensional images on some lockets and credit cards are holograms. Many characters and objects portrayed in arcade video games are also holograms. The first holographic game, "Gunsmoke," showed a three-dimensional picture of a gunslinger who shot it out with the person operating the controls. Some holograms project images into the air in front of the viewer. When the viewer reaches out to touch the image, he or she feels nothing, of course.

There are still many problems with holography. Objects that are too big cannot be photographed very well. Because monochromatic light must be used, the images produced are in one color. The only way to make multicolored pictures is by using several laser beams, each a different color, and combining them. This is very difficult to do. The images made this way do not look completely natural. Also, air molecules absorb some of the light and cause the pictures to look grainy. But scientists are working to overcome these problems.

HOLOGRAPHY

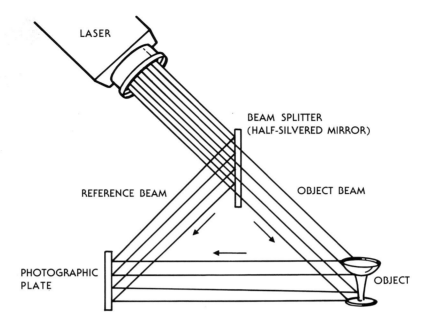

Holograms are photographs that look three-dimensional. Objects in a hologram appear to move when viewed from different angles. A hologram is made by directing a laser beam at the object to be photographed. Between the laser and the object, however, is a half-silvered mirror, or beam splitter, which splits the laser beam in two. One of the beams, called the reference beam, is reflected directly from the mirror to the photographic plate. The other, the object beam, first passes through the mirror. Then it reflects off of the object and onto the photographic plate. The interference between these two beams when they meet on the photographic plate causes the three-dimensional effect of the hologram.

Artists have tried working with holograms but, as with lasers, the equipment is expensive. So, holographic art is not yet widespread. A more practical art-related use for holography has been examining ancient paintings. When an old masterpiece is photographed to produce a hologram, experts can detect which sections of the painting are in need of repair.

Motion Picture Magic

Another use for lasers in the entertainment field is the production of special effects for movies. Several companies that produce these effects (special effects houses) use lasers in highly technical ways to help make their equipment produce truer colors. The first movie to use a laser to print images directly on the film was *Young Sherlock Holmes.* Industrial Light and Magic (ILM), perhaps the most famous special effects house, created the effects for the film. In one scene, a painted knight on a stained glass church window comes to life. The knight jumps down from the window and chases a priest out of church.

The effect was created in the following way: ILM artists painted the knight onto a TV screen using a special pen that used electricity instead of ink or paint. The image was then stored in a computer that was hooked up to the screen. Next, the artists programmed the computer to rearrange the image so it could be seen from several different angles. Then, the computer created pictures of each of the different movements the knight would make in the finished scene. When the artists ordered the computer to quickly play back all these images, the knight appeared to move around on the computer screen.

In the last and most important step, the artists transferred the computer images of the knight onto the photographic film. In creating similar effects for previous movies, this was done by simply photographing the images directly from the computer screen. But the picture on a computer screen is not as sharp and bright as filmmakers would like.

The ILM artists decided to connect the computer to a laser. The computer directed the laser to transfer or "paint" the stored images of the knight right onto the blank film. The knight now showed more detail, and the colors were much more vivid. Later, this film clip of the

In Return of the Jedi, *Luke Skywalker and Darth Vader battle each other with laser swords. The swords did not use actual lasers, however. They were drawn onto the film and illuminated with regular light.* COURTESY OF LUCASFILM LTD

moving knight was combined with separate film footage of the priest in the church. In the final version that appeared on theater screens, the knight seemed to be actually walking around inside the church.

As strange as it might sound, when moviemakers want to portray an actual laser beam on the screen, they cannot use a real laser. For instance, contrary to popular opinion, the laser swords used by Luke Skywalker and Darth Vader in the *Star Wars* films were not lasers at all. In the first place, real laser devices would not produce beams only three feet long. Instead, the beams would keep on going and punch holes in the furniture, walls, and bodies of innocent onlookers. Also,

the effect of the two beams smashing together like regular metal swords is completely imaginary. Real laser beams would just pass right through each other. This would make the fight more comical than dramatic. The most obvious problem with using real lasers (if such hand-held versions could even be built) would be the danger posed by the beams. The actors and most of the members of the film crew would all be blind within an hour.

Lasers have added a fresh, visually exciting dimension to the world of entertainment. In the years to come, it is certain that scientists and artists will continue to combine their talents to produce many inventive and dramatic new forms of laser-based entertainment.

The Future of the Laser

Modern technology is advancing so quickly that the average person simply cannot keep up with it. Even some scientists are occasionally unaware of discoveries being made in other fields. Lasers are very much a part of this technology explosion. They help scientists discover new knowledge and, as a result, fuel the explosion. At the same time, by advancing communications, lasers help spread the new knowledge to those who want it.

No one can predict what new and unheard-of discoveries will mark the next century of science. These discoveries, no doubt, will change the world in ways that cannot even be guessed at today. What can be imagined are possible ways that today's technology might be used in the near future. In the case of the laser, consider what projects are being worked on now and what other projects experts see on the immediate horizon.

A worker uses lasers to inspect power cables and make sure they are functioning properly. As laser technology advances, useful tools for business and industry will continue to be developed.

Perhaps in the future, fiber optics will connect office workers to information sources around the world. By punching a few keys, an employee may soon be able to retrieve vital data and conduct complex cross-referencing.

An Explosion of Information

Earlier, it was shown how laser fiber optics will soon connect most homes to central information storehouses. The entire contents of a library will be available on a person's home computer screen. Now, extend that idea to include the contents of all the major libraries in the world. Also include the information from museums, science labs, observatories, and government archives. Next, imagine that this huge mass of information has been cross-referenced, that is, broken down by subject. This will mean that a person who wants to know about koala bears will

simply press a button on his or her keyboard. In seconds, every known fact about koala bears will be stored in the person's computer memory. But this could amount to dozens of volumes. What if the person only wants to know what koala bears eat? Pressing another button will activate further cross-referencing. Finally, after a total of only ten or twenty seconds, the person will have fifty or sixty pages of information on the feeding habits of koala bears.

All this complex cross-referencing of information on a global scale will be a mighty task, to say the least. If done by hand, such a task would require the services of about a million people, each working twenty-four hours a day, 365 days a year for more than fifty years. And the

In schools, advances in computer and laser technologies will allow students to access libraries and other organizations that they would normally never get to visit. Students will be able to enhance their education by obtaining information that the local school library could never provide.

information bank will have to be constantly updated as new knowledge is discovered. Obviously, the only way all this will happen is through the use of a super-computer—one far more advanced than those presently operating. The optical computer discussed earlier may be the answer.

Images in Thin Air

Such laser-computer advances will also lead to the eventual perfection of holography. Experts predict that some time in the next century, three-dimensional holographic movies will become common. It will be like watching old-fashioned 3D movies, only without the special glasses. Even holographic television may be developed. This will be very difficult to construct because so much information is needed to form a holographic image. To transmit the information of a single hologram to a home, it will take a cable with the capacity of five hundred television channels. Once the hologram arrives in someone's living room, the television itself will have to be able to project the hologram. This will require a screen with more than one thousand times more detail than today's TV screens.

But many researchers believe these problems will be solved. If so, it will mean that more than entertainment can be piped into a person's home. For instance, when the phone rings, a projected image of the caller could appear in the room. The illusion will seem perfectly real, except for one thing: the person receiving the call will be able to walk right through the image of the caller.

Once perfected, this amazing technology will not be limited to telephone calls. Entire meetings will be held, in which only a few or even none of the participants are actually in the meeting room. Similarly, a teacher's hologram might be made to appear in the bedroom of a student who is home sick.

Of course, such a system could be seriously misused. For example, a disreputable government or organization could secretly plant a holograph camera in a person's home. The organization could then spy on the person by watching a completely three-dimensional hologram of his or her every movement. At the very least, this could be an embarrassing situation. It is hoped that if such technology becomes common, safeguards will be developed to keep this sort of thing from happening.

A Portable Appliance Store

It was mentioned earlier how the beam of a laser can be focused to a microscopic point. This has given the laser the ability to create discs to store vast amounts of information. Video and audio discs working on the same principle have also been described. Researchers are already talking about using this principle to miniaturize electronic devices so that they can be carried or even worn by the average person. A tiny disc programmed with billions of bits of information will become the core of each device. The devices themselves will have to be modified to work with fiber optics or some other system that eliminates conventional metal wires and circuits. This will allow the machines to be extremely small.

The end result will be a tiny unit, probably about the size of a bar of soap. The unit will include a telephone, television, radio, tape recorder and player, and a portable memory bank. Instead of using a screen, a tiny laser within the unit will project images from the television and tape player onto any blank surface the wearer desires. Such a gadget will be the real-life version of the all-purpose wrist-radio worn by superdetective Dick Tracy in the comic strips. It may be the precursor of the "tricorder" used by the characters of *Star Trek*.

Fixing Brains and Eyes

In medicine, lasers have already begun to be used in brain surgery. Usually, a laser beam burns away a tumor that has formed within the skull. The main problem with the process seems to be excess heat. Great care must be taken not to allow the hot beam to damage the brain cells surrounding the tumor. Surgeons are working to eliminate this problem.

Some researchers are hopeful that many other types of delicate brain operations will be performed by lasers. Some such treatments may use low-power laser light to cause chemical reactions in selected sections of brain tissue. These reactions might help control certain mental disorders.

The future of lasers in eye surgery already promises to bring about a world in which no one needs eyeglasses or contact lenses. Whether or not a person needs such vision aids is usually deter-

out exactly how the eye should be reshaped. The doctor will then use a different beam to cut the eyeball at various strategic points. The incisions will be welded by another beam. Eventually, the eyeball will be shaped correctly, and the person will be able to see with twenty-twenty vision.

Providing Unlimited Energy

The most ambitious and far-reaching future use for the laser will probably be the production of energy. This will be mainly in the form of electricity to power the homes, factories, offices, and machines of human civilization. Today's major methods for producing energy are water power, the burning of coal, gas, and oil, and nuclear reactors. But all these methods may not be enough to supply the energy needs of the future. The population of the world continues to grow rapidly, and more people create a demand for more energy.

Water power requires building plants near rivers or dams. There are only so many such locations, and most are not very near population centers. Supplies of coal, gas, and oil are running out, and nuclear reactors can leak radiation, creating a public danger. The laser, on the other hand, promises to open up new and seemingly endless stores of energy for humanity's use.

Production of energy by lasers will take two forms. The first is the solar-powered satellite. Plans have been considered for these devices since the 1960s, but not enough money has been spent for research and no such satellites have yet been built. However, when the government and public become interested (or desperate) enough, these devices may be produced.

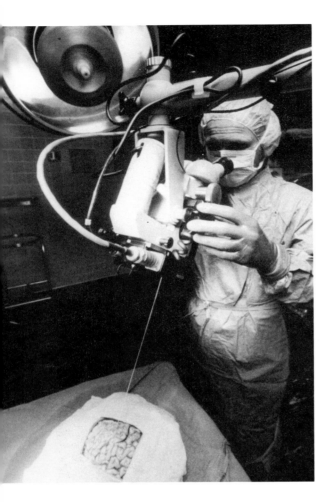

This surgeon uses a laser to perform delicate brain surgery. Because the beam can be aimed to cut very small areas, surrounding brain tissue remains untouched.

mined by the shape of the eyeball. If it is a little too short or too long, vision problems occur and corrective lenses are needed. Experiments are now taking place involving the complete reshaping of the human eye by a laser beam.

In the future, such treatments will be common. The beam will precisely measure the eye in three dimensions, pinpointing any problems. This information will be fed into a computer, which will figure

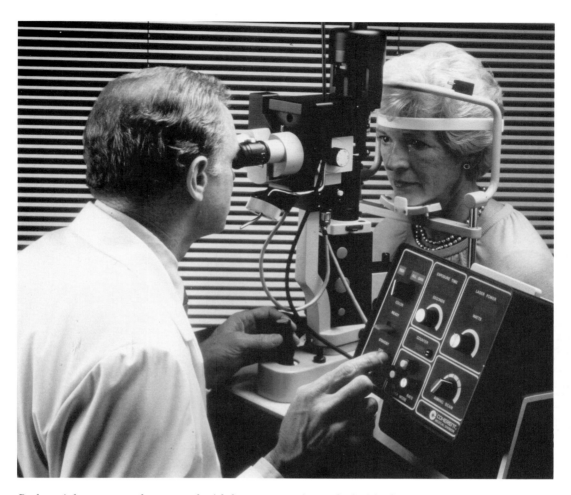

Bad eyesight may soon be corrected with laser surgery instead of with glasses. A doctor will use a laser to cut and reshape the eyeball until the patient has perfect vision.

The satellite will be rocketed into a special orbital position above the earth. Objects in this position always stay above a certain fixed point on the earth's surface. Once in position, the satellite will begin gathering energy from sunlight. The energy will power a large laser that will direct a beam back to earth. There, a receiver will collect the beam and convert it into electricity. If enough of these satellites can be put into orbit, a large share of the earth's energy needs will be met.

Some people worry that such a beam might be aimed in the wrong direction and cause death and destruction. As a matter of fact, the military has considered this method for making beam weapons. But ways will be found to adjust the power of the beam so that it will not do any damage. The time and money it will take to get these satellites orbiting will be worthwhile because sunlight is free. Since the sun is expected to shine for several billion more years, sunlight is also nearly inexhaustible.

Left: Hoover Dam, located on the Colorado River between Nevada and Arizona, is a major source of hydroelectric power. Eventually, laser-produced forms of energy may replace hydroelectric power and other traditional energy sources, such as oil and gas.

Below: Solar-powered satellites may one day gather energy from sunlight and send it to earth in a laser beam. A receiver will collect the laser beam and convert its energy to electricity.

Fusion, or the fusing of two atoms into one, releases energy in a hydrogen bomb. Lasers may one day be used to perform controlled fusion reactions that will generate heat and light.

The Power of the Atom

The other way lasers will produce energy is through nuclear fusion. This is the process that makes the sun and other stars shine. It is one of two processes that are normally referred to as atomic. The other is nuclear fission. Both of these processes have been successfully used by scientists to make atomic bombs. Fission produced the atom bomb, and fusion produced the hydrogen bomb.

Nuclear fission occurs when a particle of energy hits the center, or nucleus of an atom. The nucleus splits, sending out other particles of energy. These then hit other atoms, and the process quickly speeds up. As more and more atoms are split, a chain reaction takes place and vast amounts of energy are released in the form of heat and light. This large release of energy is what destroyed the cities of Hiroshima and Nagasaki in Japan in 1945. These bombings, which killed hundreds of thousands of people, brought an end to World War II.

Later, scientists learned how to produce fission on a smaller scale. Since they could now control the process, they called it a controlled reaction. Controlled reactions are created at nuclear power plants, where the energy produced is converted

LASERS AND NUCLEAR FUSION

Most nuclear scientists believe that in the future nuclear power will be supplied by fusion, a nuclear reaction in which two atoms are combined. But starting a fusion reaction requires an enormous initial force. Many scientists think that "laser chains" can supply that force. A laser chain consists of several laser amplifiers over a hundred feet long, which intensify the power of laser beams. The high-powered beams are directed through beam splitters and onto mirrors so that several beams strike a tiny fuel pellet from all sides at once. This causes an explosion powerful enough to trigger a fusion reaction.

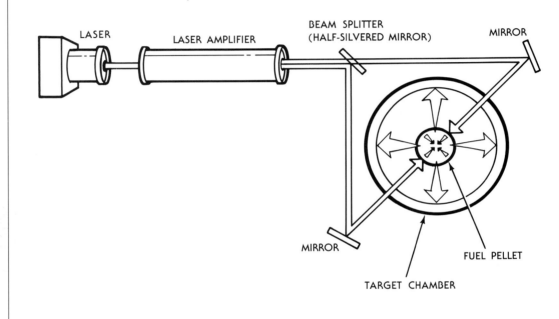

into electricity. These plants cannot blow up like a bomb. But fission only works with highly radioactive substances like plutonium. These poisonous materials can be dangerous to workers or the public in the case of a plant accident.

On the other hand, fusion occurs when two separate atoms are violently forced together. The structure of the atoms breaks down, and a new, heavier atom is formed. In the process, particles of energy are released as a by-product. In the sun and other stars, hydrogen atoms fuse to become helium atoms, and energy in the form of heat and light is released. Of course, this happens on a massive scale. In a sense, the sun is like a huge hydrogen bomb that keeps going off, second after second, year after year for billions of years.

Scientists have not been able to produce controlled fusion reactions of any consequence. This is because fusion requires a large amount of energy just to get the process going. In the sun, the trigger is the tremendous heat in the interior of the star itself. The trigger used by scientists to ignite the hydrogen bomb was an atom bomb. Obviously, scientists cannot use the same trigger to ignite controlled fusion, since the lab or power plant would be vaporized immediately in a nuclear explosion.

One great benefit of controlling fusion for energy production is that the process is relatively clean and safe. All that is needed for fuel is a small amount of hydrogen. This can be found in ordinary seawater, so the fuel is cheap and the supply almost endless. In addition, the process does not leak dangerous radiation, as does fission. The main problem with fusion has been finding a practical way to get the reaction going.

The laser may provide a way. Since the 1960s, scientists have been attempting to use laser beams to ignite nuclear fusion. So far, the experiments have succeeded only on a very small scale.

Most experts feel that large-scale controlled fusion is still many years away.

Nuclear power plants, like the Calvert Cliffs facility in Maryland, generate power by harnessing the energy that atomic reactions release. If lasers can be used to control nuclear fusion reactions, the world will have an abundant supply of safe and clean energy.

Perhaps more powerful lasers or some new technology will be needed. One thing is certain, when researchers finally succeed in igniting nuclear fusion, the world's energy problems will be solved for thousands of years to come.

A World Transformed

Less than a century has passed since Albert Einstein described the then unknown process of stimulated emission. He and other scientists at that time could not foresee the invention of the laser and its fantastic number of uses. As has happened so many times before in the history of science, the idea of a single person has grown to transform the world. This transformation will surely continue.

In the twenty-first century and beyond, the laser promises to help raise human civilization to new heights. The supertool will build a storehouse of knowledge and put that knowledge within easy reach of most people. Laser light will illuminate a complex and computerized world. This may be a world in which technology allows men and women to live increasingly productive and happy lives. Indeed, the laser may one day come to ignite the fire of the stars for humanity. The result will be clean, safe, and vastly abundant energy to fuel the engines of civilization for generations to come.

Glossary

■ ■

ammonia: A compound made up of one atom of nitrogen and three atoms of hydrogen. Ammonia gas was used in the operation of the maser.

amplify: To increase or boost the power of something.

atom: The basic unit that makes up the natural elements of which the universe is composed.

atomic bomb: A destructive weapon that uses nuclear fission as its power source.

audio disc (or compact disc): A small disc that has been encoded with music or other sounds by a laser beam. The music can be heard when the disc is inserted into a compact disc player.

bar code: Short name for the Universal Product Code, or UPC, a series of vertical bars printed on food packaging and other products in the United States. The code identifies the products when read by a helium-neon laser in the checkout lines of supermarkets.

beam weapons: Destructive laser devices.

bomb designator: A military device that uses a laser to pick out a target that will be destroyed by a bomb or missile.

cladding: The light-reflective coating on the inside lining of a fiber-optic cable.

coherent light: Light that is organized, in which all the waves are moving along in step.

collimated light: Light that is directional, in which the waves are traveling in the same direction.

core: The part of a fiber-optic cable in which the fibers themselves are located.

cornea: The transparent coating that admits light into the eye.

cycle: A wave of electromagnetic radiation (radio waves, microwaves, light, etc.).

electrical noise: Interference caused by electrical wires being placed close together.

excitation device: The power source that excites the atoms or molecules in maser and laser devices.

excited atom: An atom in which one or more electrons (particles with a negative charge) have moved temporarily to a higher energy level.

facsimile machine: A device that sends copies of photos, maps, or written documents from one place to another via electrical signals.

fiber optics: A technology in which glass fibers are used to transmit light.

frequency: The number of waves or cycles of electromagnetic radiation that pass a given point per second.

helium-neon laser: One of the most common, reliable, and least expensive lasers in use. This laser emits a deep red beam and is often used to read the UPC bar code in supermarkets.

hologram: A three-dimensional image produced by splitting coherent light into two beams and using them to expose photographic film.

holography: A special branch of photography involving the production of three-dimensional images.

hydrogen bomb: A destructive weapon that uses nuclear fusion as its power source.

laser: A device that amplifies light, producing a bright, directional, coherent beam.

laser angioplasty: Medical treatment in which plaque is cleaned from human arteries by the use of a laser beam.

lasing medium: The material that supplies the atoms or molecules that will be excited and stimulated in maser and laser devices.

LED: Light-emitting diode. A device that uses a crystal to produce a bright light.

light pipe: A hollow tube that has been lined with reflective material.

maser: A device that amplifies microwaves.

microwaves: Invisible form of electromagnetic radiation used in radar and in the maser.

modulate: To change in intensity, frequency, or brightness.

molecule: A combination of two or more atoms that forms a building block of a substance.

monochromatic: Having only one color.

nucleus: The center of an atom.

nuclear fission: Process in which atoms are split, resulting in a chain reaction and release of energy.

nuclear fusion: Process in which two separate atoms are forced together, resulting in the formation of a heavier atom and release of energy.

ophthalmology: Study of the structure and diseases of the eye.

optical computer: A computer that, if ever actually built, will use light to process information.

optical fiber: A glass fiber used in fiber optics to transmit light.

optical maser: Early name for the laser.

optical memory: Process by which information is stored on disks by laser beams.

particle theory of light: The idea that light is composed of individual particles.

patent: A document granted by a government to recognize a person as the original inventor of a device or process.

photon: The basic particle of light.

photophone: Device invented by Alexander Graham Bell in an attempt to transmit a human voice via light waves.

plaque: A tough, fatty substance that

can build up on the inside of human arteries.

radar: Device that bounces microwaves off objects to determine distance and location.

radiation (or electromagnetic radiation): Energy waves of various sizes and frequencies. Examples are radio waves, microwaves, and visible light.

range finder: Military device that uses a laser beam to determine the distance to a target.

resonator: Chamber or area in which amplification takes place in masers and lasers.

retina: Delicate membrane in the back of the eyeball that registers light.

royalty: A fee paid to an inventor or artist for use of an idea.

ruby laser: The first laser built. It uses a ruby as the lasing medium.

satellite: A man-made machine that orbits the earth.

smart bomb: A bomb or missile that uses a laser beam to home in on a target.

solar-powered satellite: A device that, when finally built, will capture energy from sunlight and transmit it down to earth. The energy will then be converted into electricity.

stimulated emission: Process described by Albert Einstein in which photons of light stimulate the production of more photons.

videodisc: A small disc that has been encoded with visual images by a laser beam. The images can be seen when the disc is inserted into a videodisc player.

wave theory of light: The idea that light is composed of waves.

For Further Reading

Sharon Begley with Kate Robbins, "Erasing Port-Wine Stains," *Newsweek*, February 27, 1989.

Alex Kozlov, "Tripping the Light Fantastic," *Discover*, December 1988.

Clifford L. Lawrence, *The Laser Book—A New Technology of Light*. New York: Prentice Hall, 1986.

Robin McKie, *Lasers*. New York: Franklin Watts, 1983.

James F. Peltz, "Bright Lights, Big Money," *Discover*, March 1988.

Science News, "Space Lasers May Benefit Blood Banks," February 13, 1988.

Thomas G. Smith, *ILM, The Art of Special Effects*. New York: Ballantine Books, 1986.

Rick Weiss, "Cleaning Cavities with a Light Touch," *Science News*, October 22, 1988.

Works Consulted

Alexandra Biesada, "Excimer Laser Surgery," *High Technology Business*, February 1989.

Alexandra Biesada, "Tooth Tech—The New Dentistry," *High Technology Business*, April 1989.

Joan Lisa Bromberg, "Birth of the Laser," *Physics Today*, October 1988.

T. Carson, "Now, Lasers Take Aim at Heart Disease," *Business Week*, December 19, 1988.

Tim Cole, "Laser-Controlled Fusion," *Popular Mechanics*, March 1989.

C. Davis, "Beyond the Blade," *Popular Mechanics*, October 1988.

Thomas F. Deutsch, "Medical Applications of Lasers," *Physics Today*, October 1988.

Charles De Vere, *Lasers, the Inside Story*. New York: Gloucester Press, 1984.

R. Hart, "What's New with Laser Printers?" *Home Office Computing*, September 1988.

Jeff Hecht and Dick Teresi, *Laser: Supertool of the 1980s*. New York: Ticknor and Fields, 1982.

High Technology Business, "Lasers Take Lead in Angioplasty Treatment," April 1989.

Gloria B. Lubkin, "Lasers Then and Now," *Physics Today*, October 1988.

Allan Maurer, *Lasers: Light Wave of the Future*. New York: Arco, 1983.

Allan Maurer, "Torch of a Thousand Suns," *Modern Maturity*, April/May 1988.

Hyrand M. Munchuryan, *Principles and Practice of Laser Technology*. Blue Ridge Summit, PA: TAB Books, 1983.

Herman Schneider, *Laser Light*. New York: McGraw-Hill, 1978.

Robert W. Seidel, "How the Military Responded to the Laser," *Physics Today*, October 1988.

J. M. Weiss, "Zap! You're Successful!" *Nation's Business*, May 1988.

Index

Apollo II, 34
atoms, 14, 17

Basov, Nikolai, 20, 21, 26
Bell, Alexander Graham, 13–14, 53
birthmarks, 63
blindness
    from diabetes, 62
    from glaucoma, 62
bomb designators, 29–30

communication systems
    use of lasers, 30–33, 51–58, 77–78
compact discs, 71
computers, 50
    and lasers, 42–44, 48–49, 74, 77–78
Corning Glass Works, 54

Def Leppard, 69

Einstein, Albert, 14–16, 18, 20, 86
entertainment industry
    and the laser, 67–79

facsimile machines, 46–47
fiber optics
    development of, 53–54
    used in communications, 55–57
Foreigner, 69
French, Norman R., 53

Gordon, James P., 18
Gould, Gordon
    work on the laser, 20–21, 26, 27, 63

Hiroshima, 83
Hockham, George, 53
holography, 72–73, 78–79
Huang, Alan, 50
Hughes Aircraft, 27, 29, 40

Huygens, Christian, 13

Industrial Light and Magic, 74

Kao, Charles, 53

lasers
    and communication systems, 30–33,
        51–58, 77–78
    and computers, 42–44, 48–49, 50, 74,
        77–78
    and energy
        nuclear fusion, 20–21, 83–86
        solar power, 80–81
    and fiber optics, 53–58
    and medicine
        as a scalpel, 59–60
        in dentistry, 64–65
        in surgery, 59–61, 79
        to clean arteries, 60–61
        to remove birthmarks, 63–64
        to repair the eye, 61–63, 79–80
    and the military, 27–33, 81
        bomb designators, 29–30
        in battle simulation, 30
        range finders, 29
        submarine communications, 30–33
        with conventional weapons, 29–30
    and printing, 43, 47–48
    and scanning, 42–50
    as expensive, 25, 67, 73
    benefits of
        economy, 28, 38, 40
        heat, 37–38, 40
        precision, 28, 29–30, 31, 34, 37, 40, 48,
            61, 65
        safety, 40, 59
        size, 40–41, 48–49, 55, 58
        speed, 28, 31, 38, 40, 48, 50, 65
    definition of, 12

lasers (*continued*)
   first working model, 22–23
   history of, 12–19, 20–23
   in entertainment
      and holography, 72–73, 78–79
      and light shows, 67–68
      and music, 68–71
      and special effects, 73–75
   in the office, 46–47
   in supermarkets, 44–45
   Nobel Prize for, 26
   operation of, 22–24
   patents on, 21, 25–26
   properties of, 12, 27–29, 31, 35–37,
      42–43, 48, 51–53
   research costs, 27, 81
   types of, 24–25, 46
      ruby, 22–23, 26, 27
   uses of, 12, 25, 67
      as a tool, 34–40
      can fight crime, 40–41
laser weapons, 28, 29, 30
light
   and fiber optics, 53–58
   in communications, 14
   particle theory of, 13
   speed of, 13
   stimulated emission of, 15, 16
   wave theory of, 13
   see also lasers

Maiman, Theodore H., 22–25, 26, 27, 40, 72
maser, 20, 22, 23
   development of, 17–19
Maurer, Robert, 54, 55
medicine
   and lasers, 59–66
microwaves
   as radiation, 16
   definition of, 16, 20

military
   use of lasers, 27–33

Nagasaki, 83
National Aeronautics and Space
      Administration (NASA), 34
Newton, Sir Isaac, 13
Nobel Prize for physics, 26
nuclear fission, 83–84
nuclear fusion, 20–21, 83–86
nuclear power plants, 84

photons, 14–15
photophone, 13–14, 53
printing
   and the lasers, 43, 47–48
Prokhorov, Aleksander, 20, 21, 26

radar, 16
radiation waves, 16, 51–52
range finders, 29
ray guns, 27
Reagan, Ronald
   Star Wars policy, 33
   inauguration of, 69
resonator, 18–19

scanning systems, 42–50
Schawlow, Arthur, 20, 21, 26
special effects
   and the laser, 73–75
Star Trek, 79
Star Wars, 75
Star Wars defense system, 33
stimulated emission of light, 15, 16, 86
   process of, 14, 17
Supermarket Institute, 44

Technical Research Group, 27
telephone signals

telephone signals (*continued*)
    and lasers, 51–52, 55–56
television signals
    and lasers, 51, 57
The Magnificent Ambersons, 72
The Who, 69
Townes, Charles
    work on the laser, 20–21, 26
    work on the maser, 16–18
Tracy, Dick, 79

U.S. Department of Defense
    funding for laser research, 27, 33
Universal Product Code, 44

videodiscs, 70–72

World War II, 16

Young Sherlock Holmes, 73

Zeiger, Herbert J., 18

About the Author

Don Nardo is an actor, makeup artist, film director, composer, and teacher, as well as a writer. As an actor, he has appeared in more than fifty stage productions, including several Shakespeare plays. He has also worked before or behind the camera in twenty films. Several of his musical compositions, including a young person's version of H.G. Wells's *The War of the Worlds,* have been played by regional orchestras. Mr. Nardo has written short stories, articles, textbooks, screenplays, and several teleplays, including an episode of ABC's "Spenser: For Hire." In addition, his screenplay *The Bet* won an award from the Massachusetts Artists Foundation. Mr. Nardo lives with his wife and son on Cape Cod, Massachusetts.

Picture Credits

■ ■